# Intelligent Systems Reference Library

Volume 142

The aim of this series is to publish a Reference Library, including novel advances and developments in all aspects of Intelligent Systems in an easily accessible and well structured form. The series includes reference works, handbooks, compendia, textbooks, well-structured monographs, dictionaries, and encyclopedias. It contains well integrated knowledge and current information in the field of Intelligent Systems. The series covers the theory, applications, and design methods of Intelligent Systems. Virtually all disciplines such as engineering, computer science, avionics, business, e-commerce, environment, healthcare, physics and life science are included. The list of topics spans all the areas of modern intelligent systems such as: Ambient intelligence, Computational intelligence, Social intelligence, Computational neuroscience, Artificial life, Virtual society, Cognitive systems, DNA and immunity-based systems, e-Learning and teaching,Human-centred computing and Machine ethics, Intelligent control, Intelligent data analysis, Knowledge-based paradigms, Knowledge management, Intelligent agents, Intelligent decision making, Intelligent network security, Interactive entertainment, Learning paradigms, Recommender systems, Robotics and Mechatronics including human-machine teaming, Self-organizing and adaptive systems, Soft computing including Neural systems, Fuzzy systems, Evolutionary computing and the Fusion of these paradigms, Perception and Vision, Web intelligence and Multimedia.

More information about this series at http://www.springer.com/series/8578

Seiki Akama · Tetsuya Murai
Yasuo Kudo

# Reasoning with Rough Sets

## Logical Approaches to Granularity-Based Framework

Springer

Seiki Akama
Kawasaki
Japan

Yasuo Kudo
Muroran Institute of Technology
Muroran
Japan

Tetsuya Murai
Chitose Institute of Science
and Technology
Chitose
Japan

ISSN 1868-4394 ISSN 1868-4408 (electronic)
Intelligent Systems Reference Library
ISBN 978-3-319-89195-8 ISBN 978-3-319-72691-5 (eBook)
https://doi.org/10.1007/978-3-319-72691-5



Printed on acid-free paper

This Springer imprint is published by Springer Nature
The registered company is Springer International Publishing AG
The registered company address is: Gewerbestrasse 11, 6330 Cham, Switzerland

# Foreword

In the last half-century, mathematical models treating various kinds of uncertainty are developed remarkably: fuzzy set theory, Dempster and Shafer's theory of evidence, rough set theory, interval analysis, imprecise probability, and so on. They are not always expressed well by the conventional probability. The issues of imprecise knowledge are acknowledged in information and computer sciences, especially in artificial intelligence, while they have been investigated for a long time by philosophers, logicians, and mathematicians. Those models are not competitive but complement one another because they treat various aspects of uncertainty.

Among those, the rough set theory is proposed in 1982 by Prof. Zdzislaw I. Pawlak. In rough set theory, objects are grouped together by their features as a homogeneous family, and certain and uncertain members of a set are identified group-wise by checking whether all members of a group belong to the set or not. This treatment of rough sets is useful in the analysis of datasets concerning classification. By checking whether all objects having same features described by attributes are classified into a same class or not, we find creditable, consistently classified data, and dubious, conflicting data. We get minimally necessary attributes by reducing attributes describing the feature of groups of objects with preserving all creditable data. We obtain rules with minimal length conditions by simplifying the description of credible data in a class without loss of the classification accuracy. In these ways, rough sets play important roles of data analysis, and most of real-world applications of rough sets utilize the attribute reduction and rule induction.

It is known that such a group-wise processing resembles human information processing. Generally, this kind of information processing is called a granular computing by Prof. Lotfi A. Zadeh, the father of fuzzy sets. In granular computing, groups can be fuzzy sets, crisp sets, intervals, and so on. Human perception is usually not numerical but categorical or ordinal. For example, let us consider temperature. We, humans, say 'hot,' 'cold,' 'comfortable,' and so on and never say '31 Celsius' or '64 Fahrenheit' until we see a thermometer. Namely, the temperature is granulated into 'hot,' 'cold,' 'comfortable,' and so on in our mind, and human perception of the temperature is categorical. Granular computing is developing toward the realization of human-like information processing such as computing with words. Rough sets provide promising tools to formulate granular computing, and further developments are expected.

The authors of this book are leading researchers in the fields of non-classical logic, rough set theory, and granular computing. By its limitation of the representability, human reasoning under uncertainty cannot be explained well by the classical logic. Non-classical logic such as modal logic, many-valued logic, intuitionistic logic, paraconsistent logic has been investigated and developed since the days of Aristotle. In this book, rough set theory is considered from viewpoints of algebras and non-classical logic. After the fundamental discussions in non-classical logic, logics for rough sets are formulated. Moreover, a granularity-based framework of reasoning, which is a general approach to reasoning with rough sets is proposed and investigated the relations to non-monotonic reasoning, association rules in conditional logic, and background knowledge.

This book is unique and would be the initial attempt that rough sets and granularity-based framework of reasoning are systematically developed from the viewpoint of non-classical logic. It is well-structured, comprehensive, and clearly written, considering the beginners and at the same time researchers in non-monotonic reasoning, rough sets, and the related topics. The book is useful and recommendable for wide range of people interested in rough sets and granular computing. No doubt this book contributes significantly and remarkably to the development of rough sets and granular computing.

Osaka, Japan
October 2017

Masahiro Inuiguchi

# Preface

Pawlak proposed rough set theory in 1982. It can be seen as an extension of set theory, in which a subset of a universe is formalized by a pair of sets, i.e., the lower and upper approximations. These approximations can be described by two operators on subsets of the universe.

In rough set theory, an equivalence relation, i.e., reflexive, symmetric, and transitive relation, plays an important role. Namely, the lower approximation of a given set is the union of all equivalence classes which are subset of the set, and the upper approximation is the union of all equivalence classes which have a non-empty intersection with the set.

The idea of a rough set has several connections with non-classical logics, in particular, modal logic. A lot of work has been done to provide a logical foundation for rough set theory. In the 1980s, a logic for reasoning about concepts, which is essentially the modal logic S5, was developed based on rough sets by Orlowska. A generalization of rough sets by modal logic using Kripke semantics was also worked out by Yao and Lin.

Now, rough set theory becomes one of the most important frameworks for imprecise and uncertain data and reasoning from data. It is also connected with granularity computing. In fact, there are many issues on various types of reasoning related to rough set theory.

This book explores reasoning with rough sets by developing some granularity-based frameworks. We begin with a brief description of rough set theory. Next, we examine some relations between rough set theory and non-classical logics including modal logic. We also develop a granularity-based framework for reasoning in which various types of reasoning can be formalized. This book will be of interest to researchers working on the areas of artificial intelligence, database, and logic.

The structure of the book is as follows.

Chapter 1 gives an introductory presentation to motivate our work on rough set theory. Rough set theory is interesting theoretically as well as practically, and a quick survey on the subject, including overview, history, and applications, is helpful to the readers.

Chapter 2 describes the foundations for rough set theory. We outline Pawlak's motivating idea and give a technical exposition. Basics of Pawlak's rough set theory and variable precision rough set model are presented with some related topics. We also present variants and related theories.

Chapter 3 surveys some non-classical logics. They are closely related to the foundations of rough set theory. We provide the basics of modal, many-valued, intuitionistic, and paraconsistent logic.

Chapter 4 introduces several logical characterizations of rough sets. We outline some approaches in the literature including double Stone algebras, Nelson algebras, and modal logics. We also discuss rough set logics, logics for reasoning about knowledge, and logics for knowledge representation.

Chapter 5 presents a granularity-based framework of deduction, induction, and abduction using variable precision rough set models proposed by Ziarko and measure-based semantics for modal logic proposed by Murai et al. This is of special importance as a general approach to reasoning based on rough set theory. We also discuss non-monotonic reasoning, association rules in conditional logic, and background knowledge.

Chapter 6 gives some conclusions with the summary of the book. We evaluate our work in connection with others. We also discuss several issues to be investigated.

We are grateful to Prof. Lakhmi C. Jain and Prof. Masahiro Inuiguchi for useful comments.

Kawasaki, Japan Seiki Akama
Chitose, Japan Tetsuya Murai
Muroran, Japan Yasuo Kudo
October 2017

# Contents

# Chapter 1
# Introduction

**Abstract** This gives an introductory presentation to motivate our work on rough set theory. Rough set theory is interesting theoretically as well as practically, and a quick survey on the subject, including overview, history and applications, is helpful to the readers.

## 1.1 Rough Set Theory

Pawlak proposed *rough set theory* in 1982; see Pawlak [1, 2]. It can be seen as an extension of (standard) *set theory*, in which a subset of a universe is formalized by a pair of sets, i.e., the lower and upper approximations. These approximations can be described by two operators on subsets of the universe.

Observe that, in rough set theory, an *equivalence relation*, i.e., reflexive, symmetric and transitive relation, plays an important role. Based on an equivalence relation, we can define the lower approximation of a given set as the union of all equivalence classes which are subset of the set, and the upper approximation as the union of all equivalence classes which have a non-empty intersection with the set. These approximations can naturally represent incomplete information.

Of course, rough set theory can be developed by relations other than an equivalence relation. But the use of an equivalence relation enables an elegant formalization, and we can obtain simple applications. However, after Pawlak's work, versions of rough set theory using various relations have been proposed in the literature.

Rough set theory is, in particular, helpful in extracting knowledge from data tables and it has been successfully applied to the areas such as data analysis, decision making, machine learning, etc.

We also observe that set theory and logic are strongly connected. This means that rough set-based approaches to knowledge representation and logic-based ones have intimate connections. In fact, *rough set* has several connections with *non-classical logics*, in particular, modal logic. A lot of work has been done to provide a logical foundation for rough set theory.

In the 1980s, a logic for reasoning about concepts, which is essentially the modal logic S5, was developed based on rough sets by Orlowska [3]. A generalization of




S. Akama et al., *Reasoning with Rough Sets*, Intelligent Systems
Reference Library 142, https://doi.org/10.1007/978-3-319-72691-5_1

rough sets by modal logic using Kripke semantics was also worked out by Yao and Lin [4].

Now, rough set theory becomes one of the most important frameworks for imprecise and uncertain data and reasoning from data. It is also connected with granular computing. In fact, there are many issues on various types of reasoning related to rough set theory.

This book explores reasoning with rough sets by developing a granularity-based framework. We begin with a brief description on rough set theory. Next, we examine some relations between rough set theory and non-classical logics including modal logic.

We also develop a granularity-based framework for reasoning in which various types of reasoning can be formalized; see Kudo, Murai and Akama [5]. This book will be of interest to researchers working on the areas in Artificial Intelligence, database and logic.

## 1.2 History

Here, we describe the history of rough set theory shortly. In 1981, Pawlak proposed *information system* in Pawlak [6]. It shares many ideas with rough set theory, and it is regarded as a forerunner of rough set theory.

In 1982, Pawlak proposed a concept of *rough set* to deal with reasoning from imprecise data in [6]. His contributions were compiled in his monograph published in 1991; see Pawlak [1].

Pawlak's starting point is to give formal classification of knowledge. Thus, rough set theory is closely related to logics for knowledge. In fact, Orlowska studied logical aspects of learning concepts in Orlowska [7] in 1988.

She proposed a logic for reasoning about knowledge in [8] in 1989. These works provided connections of rough set theory and modal logic, since her formal systems are essentially the modal logic S5.

Fariñas del Cerro and Orlowska developed *DAL*, a logic for data analysis in 1985; see [9]. *DAL* is a modal logic inspired by the ideas of rough set theory. Their work reveals that modal logic is of special interest to data analysis.

Ziarko proposed the *variable precision rough set (VPRS) models* in 1993; see Ziarko [10]. The work extends rough set theory capable of dealing with probabilistic or inconsistent information.

Yao and Lin studied the connection of general rough set model and modal logics by means of Kripke models in 1996 in Yao and Lin [4]. By their work, it became clear that lower (upper) approximation in rough sets and necessity (possibility) are closely related.

It is natural to consider the unification of rough set theory and fuzzy set theory, because both theories can handle vagueness. There are several approaches in the literature. For instance, Dubois and Prade [11] clarified the differences of these two, and proposed a *fuzzy rough set* and *rough fuzzy set* in 1989.

The former fuzzifies an equivalence relation and the latter uses upper and lower approximation on fuzzy sets. Depending on applications, we could choose one of them. Nakamura and Gao [12] also proposed a fuzzy rough set in connection with fuzzy data analysis in 1991. Their approach is based on modal logic, influenced by *DAL*.

Pagliani proposed to use *Nelson algebras* to lay an alternative foundation for rough set theory in 1996; see Pagliani [13]. Later, he discussed the roles of negations in his theory in expressing vagueness; see Pagliani [14].

A logic for rough sets was first developed by Düntsch in 1997 in Düntsch [15]. Based on Pomykala's results, he proposed a propositional logic for rough sets with an algebraic semantics based on regular double Stone algebras.

Pomykala and Pomykala [16] showed that the collection of rough sets of an approximation space forms a *regular double Stone algebra* in 1998. It is a famous fact that the collection of all subsets of a set constitutes a Boolean algebra and that its logic is exactly the classical propositional logic.

Rough set theory can serve as a semantic basis for non-classical logics. For example, Akama and Murai developed a rough set semantics for some three-valued logics in 2005; see Akama and Murai [17].

Miyamoto et al. used a family polymodal systems with the structure of lattices on the polymodal indices in [18]. They considered two applications. One is generalized possibility measures in which lattice-valued measures are proposed and relations with the ordinary possibility and necessity measures are uncovered. The other is an information system as a table such as the one in the relational database. The work generalized rough sets which are called *multi-rough sets*.

Kudo et al. proposed a granularity-based framework of deduction, induction and abduction based on VPRS models and measure-based semantics of modal logic due to Murai et al. [19–21] in 2009; see Kudo et al. [5].

Their work provides a unified formulation of various types of reasoning within the framework of rough set theory, and it can be applied to several AI problems. We will present the framework in Chap. 5.

Akama et al. proposed a Heyting-Brouwer rough set logic as an extension of Düntsch's logic for rough sets in 2013; see Akama et. al [22]. The logic is useful for reasoning about rough information, because it has an implication. It is also noted that its subsystem can be used as a logic for vagueness; see Akama et al. [23].

Rough set theory has been applied to many areas and there is in fact a rich literature. In the next section we will shortly overview on interesting applications of rough set theory. There are some textbooks on rough set theory, e.g., Polkowski [24].

## 1.3 Applications

Although the present book deals with reasoning with rough sets, we here present some examples of applications. Of course, our exposition on applications is not complete, but rough set theory can be applied to many areas and it is of special interest for engineering applications. Promising areas include, as mentioned below, machine learning, data mining, decision making, medicine, etc.

Rough set theory has an advantage over other theories in that it does not need any preliminary information or additional information about data. Namely, it dispenses with the notions like probability, basic probability assignment in Dempster-Shafer theory and membership in fuzzy set theory. The feature leads to many successful applications.

*Machine Learning* (ML) is the subfield of AI whose aim is to give computers the ability to learn. Orlowska [7] studied logical aspects of learning concepts by modal logic. A rough set approach to learning was explored by Pawlak [2] who discussed learning from examples and inductive learning.

*Data Mining* is the area which studies the process of finding patterns in large databases. It is also called *knowledge discovery in database* (KDD). Now, in the presence of Big Data, many people in various fields are investigating Data Mining. The reader should consult Adriaans and Zatinge [25] for details.

There are many different approaches to Data Mining, and rough set theory constitutes one of the helpful approaches; see Lin and Cercone [26]. This is because rough set theory can formalize information systems and the methods using decision tables are very useful for Data Mining.

*Decision making* is the process of selecting a logical choice from possible options. A system capable of doing decision making is called a *decision support system*. Decision table and its simplification methods are applied to decision-making. For a survey on the subject, see Slowinski et al. [27].

*Image processing* considers image data and their various treatments. Which is the sub-field of *pattern recognition*. Rough set theory is helpful to segmentation and extraction; see Pal et al. [28]. Other promising applications include image classification and retrieval, which challenge standard approaches.

*Switching circuit* is a basis for hardware design and some effective methods like *Karnaugh maps* have been already established. However, rough set theory can serve as an alternative base for switching circuit.

In fact, switching function can be described as a decision table and we can employ simplification methods for it. Note that the method is completely expressed in rough set theory. See Pawlak [2] for details.

*Robotics* is the area of constructing a robot. There are several types of robotics systems; from simple one to human-like one. In any case, it needs various disciplines mainly for hardware and software.

As an actual robot faces uncertainty in various stages, we can make use of rough set theory for its system. For an overview on the subject, the reader is referred to Bit and Beaubouef [29].

*Mathematics* should be re-considered in the context of rough sets. Since rough set is a generalization of standard set, rough set theory can be regarded as an extension of standard one. It may be possible to work out intriguing mathematical results in rough set theory. However, little work has been done on the subject.

*Medicine* is one of the most significant areas benefited from rough set theory, since medical data can be considered both incomplete and vague. However, physicians must diagnose a patient and decide the best way for a patient without obtaining complete information.

There are many works on medical science based on rough set theory. For instance, Tsumoto [30] proposed a model for medical diagnostic rules based on rough sets. Hirano and Tsumoto [31] applied rough set theory to the analysis of medical images.

Rough set theory gives us one of the important backgrounds of the so-called *soft computing* whose aim is to deal with inexact solutions of computational problems. Other backgrounds include fuzzy logic, evolutional computing, Machine Learning and probability theory. Rough set theory seems to have many advantages over other backgrounds.

## References

1. Pawlak, P.: Rough sets. Int. J. Comput. Inf. Sci. **11**, 341–356 (1982)
2. Pawlak, P.: Rough Sets: Theoretical Aspects of Reasoning about Data. Kluwer, Dordrecht (1991)
3. Orlowska, E.: Kripke models with relative accessibility relations and their applications to inferences from incomplete information. In: Mirkowska, G., Rasiowa, H. (eds.) Mathematical Problems in Computation Theory. pp. 327–337. Polish Scientific Publishers, Warsaw (1987)
4. Yao, Y., Lin, T.: Generalization of rough sets using modal logics. Intell. Autom. Soft Comput. **2**, 103–120 (1996)
5. Kudo, Y., Murai, T., Akama, S.: A granularity-based framework of deduction, induction, and abduction. Int. J. Approx. Reason. **50**, 1215–1226 (2009)
6. Pawlak, P.: Information systems: theoretical foundations. Inf. Syst. **6**, 205–218 (1981)
7. Orlowska, E.: Logical aspects of learning concepts. Int. J. Approx. Reason. **2**, 349–364 (1988)
8. Orlowska, E.: Logic for reasoning about knowledge. Zeitschrift für Mathematische Logik und Grundlagen der Mathematik **35**, 559–572 (1989)
9. Fariñas del Cerro, L., Orlowska, E.: DAL-a logic for data analysis. Theor. Comput. Sci. **36**, 251–264 (1985)
10. Ziarko, W.: Variable precision rough set model. J. Comput. Syst. Sci. **46**, 39–59 (1993)
11. Dubois, D., Prade, H.: Rough fuzzy sets and fuzzy rough sets. Int. J. Gen. Syst. **17**, 191–209 (1989)
12. Nakamura, A., Gao, J.: A logic for fuzzy data analysis. Fuzzy Sets Syst. **39**, 127–132 (1991)
13. Pagliani, P.: Rough sets and Nelson algebras. Fundam. Math. **27**, 205–219 (1996)
14. Pagliani, P., Intrinsic co-Heyting boundaries and information incompleteness in rough set analysis. In: Polkowski, L., Skowron, A. (eds.) Rough Sets and Current Trends in Computing. PP. 123–130. Springer, Berlin (1998)
15. Düntsch, I.: A logic for rough sets. Theor. Comput. Sci. **179**, 427–436 (1997)
16. Pomykala, J., Pomykala, J.A.: The stone algebra of rough sets. Bull. Pol. Acad. Sci. Math. **36**, 495–508 (1988)
17. Akama, S., Murai, T.: Rough set semantics for three-valued logics. In: Nakamatsu, K., Abe, J.M. (eds.) Advances in Logic Based Intelligent Systems. pp. 242–247. IOS Press, Amsterdam (2005)
18. Miyamoto, S., Murai, T., Kudo, Y.: A family of polymodal systems and its application to generalized possibility measure and multi-rough sets. JACIII **10**, 625–632 (2006)
19. Murai, T., Miyakoshi, M., Shinmbo, M.: Measure-based semantics for modal logic. In: Lowen, R., Roubens, M. (eds.) Fuzzy Logic: State of the Arts. pp. 395–405. Kluwer, Dordrecht (1993)
20. Murai, T., Miyakoshi, M., Shimbo, M.: Soundness and completeness theorems between the Dempster-Shafer theory and logic of belief. In: Proceedings of the 3rd FUZZ-IEEE on World Congress on Computational Intelligence (WCCI). pp. 855–858. (1994)

21. Murai, T., Miyakoshi, M. and Shinmbo, M.: A logical foundation of graded modal operators defined by fuzzy measures. In: Proceedings of the 4th FUZZ-IEEE, pp. 151–156. (Semantics for modal logic, Fuzzy Logic: State of the Arts, pp. 395–405, 1993) Kluwer, Dordrecht (1995)
22. Akama, S., Murai, T. and Kudo, Y.: Heyting-Brouwer Rough Set Logic. In: Proceedings of the KSE2013, Hanoi, pp. 135–145. Springer, Heidelberg (2013)
23. Akama, S., Murai, T., Kudo, Y.: Da Costa logics and vagueness. In: Proceedings of the GrC2014, Noboribetsu, Japan. (2014)
24. Polkowski, L.: Rough Sets: Mathematical Foundations. Pysica-Verlag, Berlin (2002)
25. Adsiaans, P., Zantinge, D.: Data Mining, Addison-Wesley, Reading, Mass (1996)
26. Lin, T., Cercone, N. (eds.): Rough Sets and Data Mining. Springer, Berlin (1997)
27. Slowinski, R., Greco, S., Matarazzo, B.: Rough sets and decision making. In: Meyers, R. (ed.) Encyclopedia of Complexity and Systems Science. pp. 7753–7787. Springer, Heidelberg (2009)
28. Pal, K., Shanker, B., Mitra, P.: Granular computing, rough entropy and object extraction. Pattern Recognit. Lett. **26**, 2509–2517 (2005)
29. Bit, M., Beaubouef, T.: Rough set uncertainty for robotic systems. J. Comput. Syst. Coll. **23**, 126–132 (2008)
30. Tsumoto, S.: Modelling medical diagnostic rules based on rough sets. In: Rough Sets and Current Trends in Computing. pp. 475–482. (1998)
31. Hirano, S., Tsumoto, S.: Rough representation of a region of interest in medical images. Int. J. Approx. Reason. **40**, 23–34 (2005)

# Chapter 2
# Rough Set Theory

**Abstract** This chapter describes the foundations for rough set theory. We outline Pawlak's motivating idea and give a technical exposition. Basics of Pawlak's rough set theory and variable precision rough set model are presented with some related topics. We also present variants and related theories.

## 2.1 Pawlak's Approach

We begin with an exposition of Pawlawk's approach to rough set theory based on Pawlak [1]. His motivation is to provide a theory of knowledge and classification by introducing a new concept of set, i.e. *rough set*.

By *object*, we mean anything we can think of, for example, real things, states, abstract concepts, etc.. We can assume that knowledge is based on the ability to classify objects. Thus, knowledge is necessarily connected with the variety of classification patterns related to specific parts of the real or abstract world, called the *universe of discourse* (or the universe).

Now, we turn to a formal presentation. We assume the usual notation for set theory. Let $U$ be non-empty finite set of objects called the *universe of discourse*. Any subset $X \subseteq U$ of the universe is called a *concept* or a *category* in $U$. Any family of concepts in $U$ is called *knowledge* about $U$. Note that the empty set $\emptyset$ is also a concept.

We mainly deal with concepts which form a partition (classification) of a certain universe $U$, i.e. in families $C = \{X_1, X_2, ..., X_n\}$ such that $X_i \subseteq U, X_i \neq \emptyset, X_i \cap X_j = \emptyset$ for $i \neq j, i, j = 1, ..., n$ and $\bigcup X_i = U$. A family of classifications over $U$ is called a *knowledge base* over $U$.

Classifications can be specified by using *equivalence relations*. If $R$ is an equivalence relation over $U$, then $U/R$ means the family of all equivalence classes of $R$ (or classification of $U$) referred to as categories or concepts of $R$. $[x]_R$ denotes a category in $R$ containing an element $x \in U$.

A *knowledge base* is defined as a relational system, $K = (U, \mathbf{R})$, where $U \neq \emptyset$ is a finite set called the universe, and $\mathbf{R}$ is a family of equivalence relations over $U$. $IND(K)$ means the family of all equivalence relations defined in $K$, i.e., $IND(K) = \{IND(\mathbf{P}) \mid \emptyset \neq \mathbf{P} \subseteq \mathbf{R}\}$. Thus, $IND(K)$ is the minimal set of equivalence relations,



S. Akama et al., *Reasoning with Rough Sets*, Intelligent Systems Reference Library 142, https://doi.org/10.1007/978-3-319-72691-5_2

containing all elementary relations of $K$, and closed under set-theoretical intersection of equivalence relations.

If $\mathbf{P} \subseteq \mathbf{R}$ and $\mathbf{P} \neq \emptyset$, then $\bigcap \mathbf{P}$ denotes the intersection of all equivalence relations belonging to $\mathbf{P}$, denoted $IND(\mathbf{P})$, called an *indiscernibility relation* of $\mathbf{P}$. It is also an equivalence relation, and satisfies:

$$[x]_{IND(\mathbf{P})} = \bigcap_{R \in \mathbf{P}} [x]_R.$$

Thus, the family of all equivalence classes of the equivalence relation $IND(\mathbf{P})$, i.e., $U/IND(\mathbf{P})$ denotes knowledge associated with the family of equivalence relations $\mathbf{P}$. For simplicity, we will write $U/\mathbf{P}$ instead of $U/IND(\mathbf{P})$.

$\mathbf{P}$ is also called $\mathbf{P}$*-basic knowledge*. Equivalence classes of $IND(\mathbf{P})$ are called *basic categories (concepts)* of knowledge $\mathbf{P}$. In particular, if $Q \in \mathbf{R}$, then $Q$ is called a $Q$*-elementary knowledge* (about $U$ in $K$) and equivalence classes of $Q$ are referred to as $Q$*-elementary concepts (categories)* of knowledge $\mathbf{R}$.

Now, we describe the fundamentals of rough sets. Let $X \subseteq U$ and $R$ be an equivalence relation. We say that $X$ is $R$*-definable* if $X$ is the union of some $R$-basic categories; otherwise $X$ is $R$-undefinable.

The $R$-definable sets are those subsets of the universe which can be exactly defined in the knowledge base $K$, whereas the $R$-undefinable sets cannot be defined in $K$. The $R$-definable sets are called $R$*-exact sets*, and $R$-undefinable sets are called $R$*-inexact* or $R$*-rough*.

Set $X \subseteq U$ is called *exact* in $K$ if there exists an equivalence relation $R \in IND(K)$ such that $X$ is $R$-exact, and $X$ is said to be *rough* in $K$ if $X$ is $R$-rough for any $R \in IND(K)$.

Observe that rough sets can be also defined *approximately* by using two exact sets, referred as a lower and an upper approximation of the set.

Suppose we are given knowledge base $K = (U, \mathbf{R})$. With each subset $X \subseteq U$ and an equivalence relation $R \in IND(K)$, we associate two subsets:

$$\underline{R}X = \bigcup \{Y \in U/R : Y \subseteq X\}$$
$$\overline{R}X = \bigcup \{Y \in U/R : Y \cap X \neq \emptyset\}$$

called the $R$*-lower approximation* and the $R$*-upper approximation* of $X$, respectively. They will be simply called the lower-approximation and the upper-approximation if the context is clear.

It is also possible to define the lower and upper approximation in the following two equivalent forms:

$$\underline{R}X = \{x \in U : [x]_R \subseteq X\}$$
$$\overline{R}X = \{x \in U : [x]_R \cap X \neq \emptyset\}$$

or

$$x \in \underline{R}X \text{ iff } [x]_R \subseteq X$$
$$x \in \overline{R}X \text{ iff } [x]_R \cap X \neq \emptyset.$$

The above three are interpreted as follows. The set $\underline{R}X$ is the set of all elements of $U$ which can be *certainly* classified as elements of $X$ in the knowledge $R$. The set $\overline{R}X$ is the set of elements of $U$ which can be *possibly* classified as elements of $X$ in $R$.

We define *R-positive region* ($POS_R(X)$), *R-negative region* ($NEG_R(X)$), and *R-borderline region* ($BN_R(X)$) of $X$ as follows:

$$POS_R(X) = \underline{R}X$$
$$NEG_R(X) = U - \overline{R}X$$
$$BN_R(X) = \overline{R}X - \underline{R}X.$$

The positive region $POS_R(X)$ (or the lower approximation) of $X$ is the collection of those objects which can be classified with full certainty as members of the set $X$, using knowledge $R$.

The negative region $NEG_R(X)$ is the collection of objects with which it can be determined without any ambiguity, employing knowledge $R$, that they do not belong to the set $X$, that is, they belong to the complement of $X$.

The borderline region $BN_R(X)$ is the set of elements which cannot be classified either to $X$ or to $-X$ in $R$. It is the undecidable area of the universe, i.e. none of the objects belonging to the boundary can be classified with certainty into $X$ or $-X$ as far as $R$ is concerned.

Now, we list basic formal results. Their proofs may be found in Pawlak [1]. Proposition 2.1 is obvious.

**Proposition 2.1** *The following hold:*

*(1) X is R-definable iff* $\underline{R}X = \overline{R}X$
*(2) X is rough with respect to R iff* $\underline{R}X \neq \overline{R}X$.

Proposition 2.2 shows the basic properties of approximations:

**Proposition 2.2** *The R-lower and R-upper approximations satisfy the following properties:*

(1) $\underline{R}X \subseteq X \subseteq \overline{R}X$
(2) $\underline{R}\emptyset = \overline{R}\emptyset = \emptyset, \underline{R}U = \overline{R}U = U$
(3) $\overline{R}(X \cup Y) = \overline{R}X \cup \overline{R}Y$
(4) $\underline{R}(X \cap Y) = \underline{R}X \cap \underline{R}Y$
(5) $X \subseteq Y$ *implies* $\underline{R}X \subseteq \underline{R}Y$
(6) $X \subseteq Y$ *implies* $\overline{R}X \subseteq \overline{R}Y$
(7) $\underline{R}(X \cup Y) \supseteq \underline{R}X \cup \underline{R}Y$
(8) $\overline{R}(X \cap Y) \subseteq \overline{R}X \cap \overline{R}Y$
(9) $\underline{R}(-X) = -\overline{R}X$
(10) $\overline{R}(-X) = -\underline{R}X$
(11) $\underline{R}\underline{R}X = \overline{R}\underline{R}X = \underline{R}X$
(12) $\overline{R}\overline{R}X = \underline{R}\overline{R}X = \overline{R}X$

The concept of approximations of sets can be also applied to that of membership relation. In rough set theory, since the definition of a set is associated with knowledge about the set, a membership relation must be related to the knowledge.

Then, we can define two membership relations $\underline{\in}_R$ and $\overline{\in}_R$. $x \underline{\in}_R X$ reads "$x$ *surely belongs* to $X$" and $\overline{\in}_R$ reads "$x$ *possibly belongs* to $X$". $\underline{\in}_R$ and $\overline{\in}_R$ are called the $R$-*lower membership* and $R$-*upper membership*, respectively.

Proposition 2.3 states the basic properties of membership relations:

**Proposition 2.3** *The $R$-lower and $R$-upper membership relations satisfy the following properties:*

(1) $x \underline{\in}_R X$ *imply* $x \in X$ *implies* $x \overline{\in}_R X$
(2) $X \subseteq Y$ *implies* ($x \underline{\in}_R X$ *implies* $x \underline{\in}_R Y$ *and* $x \overline{\in}_R X$ *implies* $x \overline{\in}_R Y$)
(3) $x \overline{\in}_R (X \cup Y)$ *iff* $x \overline{\in}_R X$ *or* $x \overline{\in}_R Y$
(4) $x \underline{\in}_R (X \cap Y)$ *iff* $x \underline{\in}_R X$ *and* $x \underline{\in}_R Y$
(5) $x \underline{\in}_R X$ *or* $x \underline{\in}_R Y$ *implies* $x \underline{\in}_R (X \cup Y)$
(6) $x \overline{\in}_R (X \cap Y)$ *implies* $x \overline{\in}_R X$ *and* $x \overline{\in}_R Y$
(7) $x \underline{\in}_R (-X)$ *iff non* $x \overline{\in}_R X$
(8) $x \overline{\in}_R (-X)$ *iff non* $x \underline{\in}_R X$

Approximate (rough) equality is the concept of equality in rough set theory. Three kinds of approximate equality can be introduced. Let $K = (U, \mathbf{R})$ be a knowledge base, $X, Y \subseteq U$ and $R \in IND(K)$.

(1) Sets $X$ and $Y$ are *bottom* $R$-equal ($X \bar{\sim}_R Y$) if $\underline{R}X = \underline{R}Y$
(2) Sets $X$ and $Y$ are *top* $R$-equal ($X \simeq_R Y$) if $\overline{R}X = \overline{R}Y$
(3) Sets $X$ and $Y$ are $R$-equal ($X \approx_R Y$) if $X \bar{\sim}_R Y$ and $X \simeq_R Y$

These equalities are equivalence relations for any indiscernibility relation $R$. They are interpreted as follows: $X \bar{\sim}_R Y$ means that positive example of the sets $X$ and $Y$ are the same, ($X \simeq_R Y$) means that negative example of the sets $X$ and $Y$ are the same, and ($X \approx_R Y$) means that both positive and negative examples of $X$ and $Y$ are the same.

These equalities satisfy the following proposition (we omit subscript $R$ for simplicity):

**Proposition 2.4** *For any equivalence relation, we have the following properties:*

(1) $X \bar{\sim} Y$ *iff* $X \cap X \bar{\sim} Y$ *and* $X \cap Y \bar{\sim} Y$
(2) $X \simeq Y$ *iff* $X \cup Y \simeq X$ *and* $X \cup Y \simeq Y$
(3) *If* $X \simeq X'$ *and* $Y \simeq Y'$, *then* $X \cup Y \simeq X' \cup Y'$
(4) *If* $X \bar{\sim} X'$ *and* $Y \bar{\sim} Y'$, *then* $X \cap Y \bar{\sim} X' \cap Y'$
(5) *If* $X \simeq Y$, *then* $X \cup -Y \simeq U$
(6) *If* $X \bar{\sim} Y$, *then* $X \cap -Y \bar{\sim} \emptyset$
(7) *If* $X \subseteq Y$ *and* $Y \simeq \emptyset$, *then* $X \simeq \emptyset$
(8) *If* $X \subseteq Y$ *and* $Y \simeq U$, *then* $X \simeq U$

(9) $X \simeq Y$ *iff* $-X \overline{\sim} -Y$
(10) *If* $X \overline{\sim} \emptyset$ *or* $Y \overline{\sim} \emptyset$, *then* $X \cap Y \overline{\sim} \emptyset$
(11) *If* $X \simeq U$ *or* $Y \simeq U$, *then* $X \cup Y \simeq U$.

The following proposition shows that lower and upper approximations of sets can be expressed by rough equalities:

**Proposition 2.5** *For any equivalence relation R:*

(1) $\underline{R}X$ *is the intersection of all* $Y \subseteq U$ *such that* $X \overline{\sim}_R Y$
(2) $\overline{R}X$ *is the union of all* $Y \subseteq U$ *such that* $X \simeq_R Y$.

Similarly, we can define rough inclusion of sets. It is possible to define three kinds of rough inclusions.

Let $X = (U, \mathbf{R})$ be a knowledge base, $X, Y \subseteq U$, and $R \in IND(K)$. Then, we have:

(1) Set $X$ is *bottom R-included* in $Y$ ($X \overset{\subset}{\sim}_R Y$) iff $\underline{R}X \subseteq \underline{R}Y$
(2) Set $X$ is *top R-included* in $Y$ ($X \underset{\sim}{\subset}_R Y$) iff $\overline{R}X \subseteq \overline{R}Y$
(3) Set $X$ is *R-included* in $Y$ ($X \underset{\sim}{\overset{\subset}{\sim}}_R Y$) iff $X \underset{\sim}{\subset}_R Y$ and $X \overset{\subset}{\sim}_R Y$.

Note that $\overset{\subset}{\sim}_R$, $\underset{\sim}{\subset}_R$ and $\underset{\sim}{\overset{\subset}{\sim}}_R$ are quasi ordering relations. They are called the lower, upper and rough inclusion relation, respectively. Observe that rough inclusion of sets does not imply the inclusion of sets.

The following proposition shows the properties of rough inclusion:

**Proposition 2.6** *Rough inclusion satisfies the following:*

(1) *If* $X \subseteq Y$, *then* $X \overset{\subset}{\sim} Y$, $X \underset{\sim}{\subset} Y$ *and* $X \underset{\sim}{\overset{\subset}{\sim}} Y$
(2) *If* $X \overset{\subset}{\sim} Y$ *and* $Y \overset{\subset}{\sim} X$, *then* $X \overline{\sim} Y$
(3) *If* $X \underset{\sim}{\subset} Y$ *and* $Y \underset{\sim}{\subset} X$, *then* $X \simeq Y$
(4) *If* $X \underset{\sim}{\overset{\subset}{\sim}} Y$ *and* $Y \underset{\sim}{\overset{\subset}{\sim}} X$, *then* $X \approx Y$
(5) *If* $X \underset{\sim}{\subset} Y$ *iff* $X \cup Y \simeq Y$
(6) *If* $X \overset{\subset}{\sim} Y$ *iff* $X \cap Y \overline{\sim} Y$
(7) *If* $X \subseteq Y$, $X \overline{\sim} X'$ *and* $Y \overline{\sim} Y'$, *then* $X' \overset{\subset}{\sim} Y'$
(8) *If* $X \subseteq Y$, $X \simeq X'$ *and* $Y \simeq Y'$, *then* $X' \underset{\sim}{\subset} Y'$
(9) *If* $X \subseteq Y$, $X \approx X'$ *and* $Y \approx Y'$, *then* $X' \underset{\sim}{\overset{\subset}{\sim}} Y'$
(10) *If* $X' \underset{\sim}{\subset} X$ *and* $Y' \underset{\sim}{\subset} Y$, *then* $X' \cup Y' \underset{\sim}{\subset} X \cup Y$
(11) $X' \overset{\subset}{\sim} X$ *and* $Y' \overset{\subset}{\sim} Y$, *then* $X' \cap Y' \overset{\subset}{\sim} X \cap Y$
(12) $X \cap Y \overset{\subset}{\sim} Y \underset{\sim}{\subset} X \cup Y$
(13) *If* $X \overset{\subset}{\sim} Y$ *and* $X \overline{\sim} Z$, *then* $Z \overset{\subset}{\sim} Y$

*(14) If* $X \,\tilde{\subset}\, Y$ *and* $X \simeq Z$, *then* $Z \,\tilde{\subset}\, Y$

*(15) If* $X \underset{\sim}{\tilde{\subset}} Y$ *and* $X \approx Z$, *then* $Z \underset{\sim}{\tilde{\subset}} Y$

The above properties are not valid if we replace $\overline{\sim}$ by $\simeq$ (or conversely). If $R$ is an equivalence relation, then all three inclusions reduce to ordinary inclusion.

## 2.2 Variable Precision Rough Set Models

Ziarko generalized Pawlak's original rough set model in Ziarko [2], which is called the *variable precision rough set model* (VPRS model) to overcome the inability to model uncertain information, and is directly derived from the original model without any additional assumptions.

As the limitations of Pawlak's rough set model, Ziarko discussed two points. One is that it cannot provide a classification with a controlled degree of uncertainty. Some level of uncertainty in the classification process gives a deeper or better understanding for data analysis.

The other is that the original model has the assumption that the universe $U$ of data objects is known. Therefore, all conclusions derived from the model are applicable only to this set of objects. It is useful to introduce uncertain hypotheses about properties of a larger universe.

Ziarko's extended rough set model generalizes the standard set inclusion relation, capable of allowing for some degree of misclassification in the largely correct classification.

Let $X$ and $Y$ be non-empty subsets of a finite universe $U$. $X$ is included in $Y$, denoted $Y \supseteq X$, if for all $e \in X$ implies $e \in Y$. Here, we introduce the measure $c(X, Y)$ of the relative degree of misclassification of the set $X$ with respect to set $Y$ defined as:

$$c(X, Y) = 1 - \text{card}(X \cap Y)/\text{card}(X) \text{ if } \text{card}(X) > 0 \text{ or}$$
$$c(X, Y) = 0 \text{ if } \text{card}(X) = 0$$

where card denotes set cardinality.

The quantity $c(X, Y)$ will be referred to as the relative classification error. The actual number of misclassification is given by the product $c(X, Y) * \text{card}(X)$ which is referred to as an absolute classification error.

We can define the inclusion relationship between $X$ and $Y$ without explicitly using a general quantifier:

$$X \subseteq Y \text{ iff } c(X, Y) = 0$$

The *majority* requirement implies that more than 50% of $X$ elements should be in common with $Y$. The *specified majority* requirement imposes an additional requirement. The number of elements of $X$ in common with $Y$ should be above 50% and not below a certain limit, e.g. 85%.

According to the specified majority requirement, the admissible classification error $\beta$ must be within the range $0 \leq \beta < 0.5$. Then, we can define the majority inclusion relation based on this assumption.

$$X \overset{\beta}{\subseteq} Y \text{ iff } c(X, Y) \leq \beta$$

The above definition covers the whole family of $\beta$-majority relation. However, the majority inclusion relation does not have the transitivity relation.

The following two propositions indicate some useful properties of the majority inclusion relation:

**Proposition 2.7** *If* $A \cap B = \emptyset$ *and* $B \overset{\beta}{\supseteq} X$, *then it is not true that* $A \overset{\beta}{\supseteq} X$.

**Proposition 2.8** *If* $\beta_1 < \beta_2$, *then* $Y \overset{\beta_1}{\supseteq} X$ *implies* $Y \overset{\beta_2}{\supseteq} X$.

For the VPRS-model, we define the approximation space as a pair $A = (U, R)$, where $U$ is a non-empty finite universe and $R$ is the equivalence relation on $U$. The equivalence relation $R$, referred to as an indiscernibility relation, corresponds to a partitioning of the universe $U$ into a collection of equivalence classes or elementary set $R^* = \{E_1, E_2, ..., E_n\}$.

Using a majority inclusion relation instead of the inclusion relation, we can obtain generalized notions of $\beta$-lower approximation (or $\beta$-positive region $\text{POSR}_\beta(X)$) of the set $U \supseteq X$:

$$\underline{R}_\beta X = \bigcup \{E \in R^* : X \overset{\beta}{\supseteq} E\} \text{ or, equivalently,}$$

$$\underline{R}_\beta X = \bigcup \{E \in R^* : c(E, X) \leq \beta\}$$

The $\beta$-upper approximation of the set $U \supseteq X$ can be also defined as follows:

$$\overline{R}_\beta X = \bigcup \{E \in R^* : c(E, X) < 1 - \beta\}$$

The $\beta$-boundary region of a set is given by

$$\text{BNR}_\beta X = \bigcup \{E \in R^* : \beta < c(E, X) < 1 - \beta\}.$$

The $\beta$-negative region of $X$ is defined as a complement of the $\beta$-upper approximation:

$$\text{NEGR}_\beta X = \bigcup \{E \in R^* : c(E, X) \geq 1 - \beta\}.$$

The lower approximation of the set $X$ can be interpreted as the collection of all those elements of $U$ which can be classified into $X$ with the classification error not greater than $\beta$.

The $\beta$-negative region of $X$ is the collection of all those elements of $U$ which can be classified into the complement of $X$, with the classification error not greater than $\beta$. The latter interpretation follows from Proposition 2.9:

**Proposition 2.9** *For every $X \subseteq Y$, the following relationship is satisfied:*

$$\mathrm{POSR}_\beta(-X) = \mathrm{NEGR}_\beta X.$$

The $\beta$-boundary region of $X$ consists of all those elements of $U$ which cannot be classified either into $X$ or into $-X$ with the classification error not greater than $\beta$.

Notice here that the law of excluded middle, i.e. $p \vee \neg p$, where $\neg p$ is the negation of $p$, holds in general for imprecisely specified sets.

Finally, the $\beta$-upper approximation $\overline{R}_\beta X$ of $X$ includes all those elements of $U$ which cannot be classified into $-X$ with the error not greater than $\beta$. If $\beta = 0$ then the original rough set model is a special case of VPRS-model, as the following proposition states:

**Proposition 2.10** *Let $X$ be an arbitrary subset of the universe $U$:*

*(1)* $\underline{R}_0 X = \underline{R}X$, *where $\underline{R}X$ is a lower approximation defined as* $\underline{R}X = \bigcup\{E \in R^* : X \supseteq E\}$

*(2)* $\overline{R}_0 X = \overline{R}X$, *where $\overline{R}X$ is an upper approximation defined as* $\overline{R}X = \bigcup\{E \in R^* : E \cap X \neq \emptyset\}$

*(3)* $\mathrm{BNR}_0 X = \mathrm{BN}_R X$, *where $\mathrm{BN}_R X$ is the set $X$ boundary region defined as* $\mathrm{BN}_R X = \overline{R}X - \underline{R}X$

*(4)* $\mathrm{NEGR}_0 X = \mathrm{NEG}_R X$, *where $\mathrm{NEG}_R X$ is the set $X$ negative region defined as* $\mathrm{NEG}_R X = U - \overline{R}X$

In addition, we have the following proposition:

**Proposition 2.11** *If $0 \leq \beta < 0.5$ then the properties listed in Proposition 2.10 and the following are also satisfied:*

$$\underline{R}_\beta X \supseteq \underline{R}X,$$
$$\overline{R}X \supseteq \overline{R}_\beta X,$$
$$\mathrm{BN}_R X \supseteq \mathrm{BNR}_\beta X,$$
$$\mathrm{NEGR}_\beta X \supseteq \mathrm{NEG}_R X.$$

Intuitively, with the decrease of the classification error $\beta$ the size of the positive and negative regions of $X$ will shrink, whereas the size of the boundary region will grow.

With the reduction of $\beta$ fewer elementary sets will satisfy the criterion for inclusion in $\beta$-positive or $\beta$negative regions. Thus, the size of the boundary will increase.

The reverse process can be done with the increase of $\beta$.

**Proposition 2.12** *With the $\beta$ approaching the limit* 0.5, *i.e.,* $\beta \to 0.5$, *we obtain the following:*

$$\underline{R}_\beta X \to \underline{R}_{0.5} X = \bigcup \{E \in R^* : c(E, X) < 0.5\},$$
$$\overline{R}_\beta X \to \overline{R}_{0.5} X = \bigcup \{E \in R^* : c(E, X) \leq 0.5\},$$
$$\mathrm{BNR}_\beta X \to \mathrm{BNR}_{0.5} X = \bigcup \{E \in R^* : c(E, X) = 0.5\},$$
$$\mathrm{NEGR}_\beta X \to \mathrm{NEGR}_{0.5} X = \bigcup \{E \in R^* : c(E, X) > 0.5\}.$$

The set $\mathrm{BNR}_{0.5} X$ is called an *absolute boundary* of $X$ because it is included in every other boundary region of $X$.

The following Proposition 2.13 summarizes the primary relationships between set $X$ discernibility regions computed on 0.5 accuracy level and higher levels.

**Proposition 2.13** *For boundary regions of X, the following hold:*

$$\mathrm{BRN}_{0.5} X = \bigcap_\beta \mathrm{BNR}_\beta X,$$

$$\overline{R}_{0.5} X = \bigcap_\beta \overline{R}_\beta X,$$

$$\underline{R}_{0.5} X = \bigcup_\beta \underline{R}_\beta X,$$

$$\mathrm{NEGR}_{0.5} X = \bigcup_\beta \mathrm{NEGR}_\beta X.$$

The absolute boundary is very "narrow", consisting only of those sets which have 50/50 aplite of elements among set $X$ interior and its exterior. All other elementary sets are classified either into positive region $\underline{R}_{0.5} X$ or the negative region $\mathrm{NEGR}_{0.5} X$.

We turn to the measure of approximation. To express the degree with which a set $X$ can be approximately characterized by means of elementary sets of the approximation space $A = (U, R)$, we will generalize the accuracy measure introduced in Pawlak [3].

The $\beta$-accuracy for $0 \leq \beta < 0.5$ is defined as

$$\alpha(R, \beta, X) = \mathrm{card}(\underline{R}_\beta X) / \mathrm{card}(\overline{R}_\beta X).$$

The $\beta$-accuracy represents the imprecision of the approximate characterization of the set $X$ relative to assumed classification error $\beta$.

Note that with the increase of $\beta$ the cardinality of the $\beta$-upper approximation will tend downward and the size of the $\beta$-lower approximation will tend upward which leads to the conclusion that is consistent with intuition that relative accuracy may increase at the expense of a higher classification error.

The notion of discernibility of set boundaries is relative. If a large classification error is allowed then the set $X$ can be highly discernable within assumed classification limits. When smaller values of the classification tolerance are assumed in may become more difficult to discern positive and negative regions of the set to meet the narrow tolerance limits.

The set $X$ is said to be *$\beta$-discernable* if its $\beta$-boundary region is empty or, equivalently, if

$$\underline{R}_\beta X = \overline{R}_\beta X.$$

For the $\beta$-discernable sets the relative accuracy $\alpha(R, \beta, X)$ is equal to unity. The discernable status of a set change depending on the value of $\beta$. In general, the following properties hold:

**Proposition 2.14** *If X is discernable on the classification error level* $0 \leq \beta < 0.5$, *then X is also discernable at any level* $\beta_1 > \beta$.

**Proposition 2.15** *If* $\overline{R}_{0.5}X \neq \underline{R}_{0.5}X$, *then X is not discernable on every classification error level* $0 \leq \beta < 0.5$.

Proposition 2.16 emphasizes that a set with a non-empty absolute boundary can never be discerned. In general, one can easily demonstrate the following:

**Proposition 2.16** *If X is not discernable on the classification error level* $0 \leq \beta < 0.5$, *then X is also not discernible at any level* $\beta_1 < \beta$.

Any set $X$ which is not discernable for every $\beta$ is called indiscernible or absolutely rough. The set $X$ is absolutely rough iff $\mathrm{BNR}_{0.5}X \neq \emptyset$. Any set which is not absolutely rough will be referred to as relatively rough or weakly discernable.

For each relatively rough set $X$, there exists such a classification error level $\beta$ that $X$ is discernable on this level.

Let $\mathrm{NDIS}(R, X) = \{0 \leq \beta < 0.5 : \mathrm{BNR}_\beta(X) \neq \emptyset\}$.Then, $\mathrm{NDIS}(R, X)$ is a range of all those $\beta$ values for which $X$ is indiscernible.

The least value of classification error $\beta$ which makes $X$ discernable will be referred to as discernibility threshold. The value of the threshold is equal to the least upper bound $\zeta(R, X)$ of $\mathrm{NDIS}(X)$, i.e.,

$$\zeta(R, X) = \sup \mathrm{NDIS}(R, X).$$

Proposition 2.17 states a simple property which can be used to find the discernibility threshold of a weakly discernible set $X$:

**Proposition 2.17** $\zeta(R, X) = \max(m_1, m_2)$, *where*

$$m_1 = 1 - \min\{c(E, X) : E \in R^* \text{ and } 0.5 < c(E, X)\},$$
$$m_2 = \max\{c(E, X) : E \in R^* \text{ and } c(E, X) < 0.5\}.$$

The discernibility threshold of the set $X$ equals a minimal classification error $\beta$ which can be allowed to make this set $\beta$-discernible.

We give some fundamental properties of $\beta$-approximations.

**Proposition 2.18** *For every* $0 \leq \beta < 0.5$, *the following hold:*

*(1a)* $X \overset{\beta}{\supseteq} \underline{R}_\beta X$
*(1b)* $\overline{R}_\beta X \supseteq \underline{R}_\beta X$
*(2)* $\underline{R}_\beta \emptyset = \overline{R}_\beta \emptyset = \emptyset$; $\underline{R}_\beta U = \overline{R}_\beta U = U$
*(3)* $\overline{R}_\beta (X \cup Y) \supseteq \overline{R}_\beta X \cup \overline{R}_\beta Y$

(4) $\underline{R}_{\beta}X \cap \underline{R}_{\beta}Y \supseteq \underline{R}_{\beta}(X \cap Y)$
(5) $\underline{R}_{\beta}(X \cup Y) \supseteq \underline{R}_{\beta} \cup \underline{R}_{\beta}Y$
(6) $\overline{R}_{\beta}X \cap \overline{R}_{\beta}Y \supseteq \overline{R}_{\beta}(X \cap Y)$
(7) $\underline{R}_{\beta}(-X) = -\overline{R}_{\beta}(X)$
(8) $\overline{R}_{\beta}(-X) = -\underline{R}_{\beta}(X)$

We finish the outline of variable precision rough set model, which can be regarded as a direct generalization of the original rough set model. Consult Ziarko [2] for more details. As we will be discussed later, it plays an important role in our approach to rough set based reasoning.

Shen and Wang [4] proposed the VPRS model over two universes using inclusion degree. They introduced the concepts of the reverse lower and upper approximation operators and studied their properties. They introduced the approximation operators with two parameters as a generalization of the VPRS-model over two universes.

## 2.3 Related Theories

There are many related theories which extend the original rough set theory in various aspects. Very interesting are theories which integrate both rough set theory and fuzzy set theory. In this section, we briefly review some of such theories.

Before describing related fuzzy-based rough set theories, we need to give a concise exposition of *fuzzy set theory*, although there are many possible descriptions in the literature.

*Fuzzy set* was proposed by Zadeh [5] to model fuzzy concepts, which cannot be formalized in classical set theory. Zadeh also developed a *theory of possibility* based on fuzzy set theory in Zadeh [6]. In fact, fuzzy set theory found many applications in various areas.

Let $\mathscr{U}$ be a set. Then, a fuzzy set is defined as follows:

**Definition 2.1** (*Fuzzy set*) A *fuzzy set* of $\mathscr{U}$ is a function $u : \mathscr{U} \to [0, 1]$. $\mathscr{F}_{\mathscr{U}}$ will denote the set of all fuzzy sets of $\mathscr{U}$.

Several operations on fuzzy sets are defined as follows:

**Definition 2.2** For all $u, v \in \mathscr{F}_{\mathscr{U}}$ and $x \in \mathscr{U}$, we put

$(u \vee v)(x) = \sup\{u(x), v(x)\}$
$(u \wedge v)(x) = \inf\{u(x), v(x)\}$
$\overline{u}(x) = 1 - u(x)$

**Definition 2.3** Two fuzzy sets $u, v \in \mathscr{F}_{\mathscr{U}}$ are said to be *equal* iff for every $x \in \mathscr{U}$, $u(x) = v(x)$.

**Definition 2.4** $\mathbf{1}_{\mathscr{U}}$ and $\mathbf{0}_{\mathscr{U}}$ are the fuzzy sets of $\mathscr{U}$ such that for all $x \in \mathscr{U}$, $\mathbf{1}_{\mathscr{U}} = 1$ and $\mathbf{0}_{\mathscr{U}} = 0$

It is easy to prove that $\langle \mathscr{F}_{\mathscr{U}}, \wedge, \vee \rangle$ is a complete lattice having infinite distributive property. Furthermore, $\langle \mathscr{F}_{\mathscr{U}}, \wedge, \vee, ^{-} \rangle$ constitutes an algebra, which in general is not Boolean (for details, see Negoita and Ralescu [7]).

Since both rough set theory and fuzzy set theory aim to formalize related notions, it is natural to integrate these two theories. In 1990, Dubois and Prade introduced *fuzzy rough sets* as a fuzzy generalization of rough sets.

They considered two types of generalizations. One is the upper and lower approximation of a fuzzy set, i.e., *rough fuzzy set*. The other provided an idea of turning the equivalence relation into a fuzzy similarity relation, yielding a *fuzzy rough set*.

Nakamura and Gao [8] also studied fuzzy rough sets and developed a logic for fuzzy data analysis. Their logic can be interpreted as a modal logic based on fuzzy relations. They related similarity relation on a set of objects to rough sets.

Quafafou [9] proposed *$\alpha$-rough set theory* ($\alpha$-RST) in 2000. In $\alpha$-RST, all basic concepts of rough set theory are generalized. He described approximations of fuzzy concepts and their properties.

In addition, in $\alpha$-RST, the notion of $\alpha$-dependency, i.e., a set of attributes which depends on another with a given degree in [0, 1]., is introduced. It can be seen as a partial dependency. Note that $\alpha$-RST has a feature of the ability of the control of the universe partitioning and the approximation of concepts.

Cornelis et al. [10] proposed *intuitionistic fuzzy rough sets* to describe incomplete aspects of knowledge based on *intuitionistic fuzzy sets* due to Atanassov [11] in 2003. Their approach adopted the idea that fuzzy rough sets should be intuitionistic.[1]

These works enhance the power of rough set theory by introducing fuzzy concepts in various ways. Fuzzy rough sets are more useful than the original rough sets, and they can be applied to more complicated problems.

## 2.4 Formal Concept Analysis

As a different area, *formal concept analysis* (FCA) has been developed; see Ganter and Wille [12]. It is based on *concept lattice* to model relations of concept in a precise way. Obviously, rough set theory and formal concept analysis share similar idea. We here present the outline of formal concept analysis in some detail.

FCA uses the notion of a formal concept as a mathematical formulation of the notion of a concept in Port-Royal logic. According to Port-Royal, a concept is determined by a collection of objects, called an *extent* which fall under the concept and a collection of attributes called an *intent* covered by the concepts. Concepts are ordered by a subconcept-superconcept relation which is based on inclusion relation on objects and attributes. We formally define these notions used in FCA.

[1]By intuitionistic, it means that the law of excluded middle fails. However, this does not always mean rough sets founded on the so-called intuitionistic logic.

A *formal context* is a triplet $\langle X, Y, I \rangle$, where $X$ and $Y$ are non-empty set, and $I$ is a binary relation, i.e., $I \subseteq X \times Y$. Elements $x$ from $X$ are called *objects*, elements $y$ from $Y$ are called *attributes*, $\langle x, y \rangle \in I$ indicates that $x$ has attribute $y$.

For a given cross-table with $n$ rows and $m$ columns, a corresponding formal context $\langle X, Y, I \rangle$ consists of a set $X = \{x_1, ..., x_n\}$, a set $Y = \{y_1, ..., y_m\}$, and a relation $I$ defined by $\langle x_i, y_j \rangle \in I$ iff the table entry corresponding to row $i$ and column $j$ contains $\times$.

Concept-forming operators are defined for every formal context. For a formal context $\langle X, Y, I \rangle$, operators $^{\uparrow} : 2^X \to 2^Y$ and $^{\downarrow} : 2^Y \to 2^X$ are defined for every $A \subseteq X$ and $B \subseteq Y$ by

$$A^{\uparrow} = \{y \in Y \mid for\ each\ x \in A : \langle x, y \rangle \in I\}$$
$$B^{\downarrow} = \{x \in X \mid for\ each\ y \in B : \langle x, y \rangle \in I\}$$

Formal concepts are particular clusters in cross-tables, defined by means of attribute sharing. A *formal concept* in $\langle X, Y, I \rangle$ is a pair $\langle A, B \rangle$ of $A \subseteq X$ and $B \subseteq Y$ such that $A^{\uparrow} = B$ and $B^{\downarrow} = A$.

It is noticed that $\langle A, B \rangle$ is a formal concept iff $A$ contains just objects sharing all attributes from $B$ and $B$ contains just attributes shared by all objects from $A$. Thus, mathematically $\langle A, B \rangle$ is a formal concept iff $\langle A, B \rangle$ is a fixpoint of the pair $\langle ^{\uparrow}, ^{\downarrow} \rangle$ of the concept-forming operators.

Consider the following table:

| $I$ | $y_1$ | $y_2$ | $y_3$ | $y_4$ |
|---|---|---|---|---|
| $x_1$ | × | × | × | × |
| $x_2$ | × | | × | × |
| $x_3$ | | × | × | × |
| $x_4$ | | × | × | × |
| $x_5$ | × | | | |

Here, formal concept

$$\langle A_1, B_1 \rangle = \langle \{x_1, x_2, x_3, x_4\}, \{y_3, y_4\} \rangle$$

because

$$\{x_1, x_2, x_3, x_4\}^{\uparrow} = \{y_3, y_4\}$$
$$\{y_3, y_4\}^{\downarrow} = \{x_1, x_2, x_3, x_4\}$$

Here, the following relationships hold:

$$\{x_2\}^{\uparrow} = \{y_1, y_3, y_4\}, \{x_2, x_3\}^{\uparrow} = \{y_3, y_4\}$$
$$\{x_1, x_4, x_5\}^{\uparrow} = \emptyset$$
$$X^{\uparrow} = \emptyset, \emptyset^{\uparrow} = Y$$
$$\{y_1\}^{\downarrow} = \{x_1, x_2, x_5\}, \{y_1, y_2\}^{\downarrow} = \{x_1\}$$
$$\{y_2, y_3\}^{\downarrow} = \{x_1, x_3, x_4\}, \{y_2, y_3, y_4\}^{\downarrow} = \{x_1, x_3, x_4\}$$
$$\emptyset^{\downarrow} = X, Y^{\downarrow} = \{x_1\}$$

Concepts are naturally ordered by a subconcept-superconcept relation. The subconcept-superconcept relation, denoted $\leq$, is based on inclusion relation on objects and attributes. For formal concepts $\langle A_1, B_1 \rangle$ and $\langle A_2, B_2 \rangle$ of $\langle A, Y, I \rangle$, $\langle A_1, B_1 \rangle \leq \langle A_2, B_2 \rangle$ iff $A_1 \subseteq A_2$ (iff $B_2 \subseteq B_1$).

In the above example, the following hold:

$\langle A_1, B_1 \rangle = \{\{x_1, x_2, x_3, x_4\}, \{y_3, y_4\}\}$

$\langle A_2, B_2 \rangle = \{\{x_1, x_3, x_4\}, \{y_2, y_3, y_4\}\}$

$\langle A_3, B_3 \rangle = \{\{x_1, x_2\}, \{y_1, y_3, y_4\}\}$

$\langle A_4, B_4 \rangle = \langle \{x_1, x_2, x_5\}, \{y_1\} \rangle$

$\langle A_3, B_3 \rangle \leq \langle A_1, B_1 \rangle$

$\langle A_3, B_3 \rangle \leq \langle A_4, B_4 \rangle$

$\langle A_2, B_2 \rangle \leq \langle A_1, B_1 \rangle$

$\langle A_1, B_1 \rangle \| \langle A_4, B_4 \rangle$ (incomparable)

$\langle A_2, B_2 \rangle \| \langle A_4, B_4 \rangle$

$\langle A_3, B_3 \rangle \| \langle A_2, B_2 \rangle$

We denote by $\mathscr{B}(X, Y, I)$ the collection of all formal concepts of $\langle X, Y, I \rangle$, i.e.,

$$\mathscr{B}(X, Y, I) = \{\langle A, B \rangle \in 2^X \times 2^X \mid A^\uparrow = B, B^\downarrow = A\}.$$

$\mathscr{B}(X, Y, I)$ equipped with the subconcept-superconcept ordering $\leq$ is called a *concept lattice* of $\langle X, Y, I \rangle$. $\mathscr{B}(X, Y, I)$ represents all clusters which are hidden in data $\langle X, Y, I \rangle$. We can see that $\langle \mathscr{B}(X, Y, I), \leq \rangle$ is a lattice.

Extents and intents of concepts are defined as follows:

$\text{Ext}(X, Y, I) = \{A \in 2^X \mid \langle A, B \rangle \in \mathscr{B}(X, Y, I)\ for\ some\ B\}$ (extent of concepts)

$\text{Int}(X, Y, I) = \{A \in 2^Y \mid \langle A, B \rangle \in \mathscr{B}(X, Y, I)\ for\ some\ A\}$ (intent of concepts)

Formal concepts can be also defined as maximal rectangles in the cross-table: A rectangle in $\langle X, Y, I \rangle$ is a pair $\langle A, B \rangle$ such that $A \times B \subseteq I$, i.e., for each $x \in A$ and $y \in B$ we have $\langle x, y \rangle \in I$. For rectangles $\langle A_1, B_1 \rangle$ and $\langle A_2, B_2 \rangle$, put $\langle A_1, B_1 \rangle \sqsubseteq \langle A_2, B_2 \rangle$ iff $A_1 \subseteq A_2$ and $B_1 \subseteq B_2$.

We can prove that $\langle A, B \rangle$ is a formal concept of $\langle X, Y, I \rangle$ iff $\langle A, B \rangle$ is a maximal rectangle in $\langle X, Y, I \rangle$. Consider the following table.

| $I$ | $y_1$ | $y_2$ | $y_3$ | $y_4$ |
|---|---|---|---|---|
| $x_1$ | × | × | × | × |
| $x_2$ | × | | × | × |
| $x_3$ | | × | × | × |
| $x_4$ | | × | × | × |
| $x_5$ | × | | | |

In this table, $\langle\{x_1, x_2, x_3\}, \{y_3, y_4\}\rangle$ is a rectangle which is not maximal with respect to $\sqsubseteq$. $\langle\{x_1, x_2, x_3, x_4\}, \{y_3, y_4\}\rangle$ is a rectangle which is maximal with respect to $\sqsubseteq$. The notion of rectangle can server as a basis for geometrical reasoning in formal concept analysis.

There are two basic mathematical structures behind formal concept analysis, i.e., Galois connections (cf. Ore [13]) and closure operators. A *Galois connection* between sets $X$ and $Y$ is a pair $\langle f, g\rangle$ of $f : 2^X \to 2^Y$ and $g : 2^Y \to 2^X$ satisfying $A, A_1, A_2, B, B_1, B_2 \subseteq Y$:

$$A_1 \subseteq A_2 \Rightarrow f(A_2) \subseteq f(A_1)$$
$$B_1 \subseteq B_2 \Rightarrow g(B_2) \subseteq f(B_1)$$
$$A \subseteq g(f(A))$$
$$B \subseteq f(g(B)).$$

For a Galois connection $\langle f, g\rangle$ between sets $X$ and $Y$, the set:

$$\text{fix}(\langle f, g\rangle) = \{\langle A, B\rangle \in 2^X \times 2^X \mid f(A) = B, g(B) = A\}$$

is called a set of *fixpoint* of $\langle f, g\rangle$.

Here, we show a basic property of concept-forming operators. That is, for a formal context $\langle X, Y, I\rangle$, the pair $\langle {}^{\uparrow_I}, {}^{\downarrow_I}\rangle$ of operators induced by $\langle X, Y, I\rangle$ is a Galois connection between $X$ and $Y$.

As consequence of the property, it is shown that for a Galois connection $\langle f, g\rangle$ between $X$ and $Y$, $f(A) = f(g(f(A)))$ and $g(B) = g(f(g(B)))$ for any $A \subseteq X$ and $B \subseteq Y$.

*Closure operators* result from the concept-forming operators by their composition. If $\langle f, g\rangle$ is a Galois connection between $X$ and $Y$, then $C_X = g \circ f$ is a closure operator on $X$ and $C_Y = f \circ g$ is a closure operator on $Y$.

We can show that extents and intents are just the images under the concept-forming operators as follows:

$$\text{Ext}(X, Y, I) = \{B^{\downarrow} \mid B \subseteq Y\}$$
$$\text{Int}(X, Y, I) = \{A^{\uparrow} \mid A \subseteq X\}.$$

The following relationships hold for any formal context $\langle X, Y, I\rangle$:

$$\text{Ext}(X, Y, I) = \text{fix}({}^{\uparrow\downarrow})$$
$$\text{Int}(X, Y, I) = \text{fix}({}^{\downarrow\uparrow})$$
$$\mathscr{B}(X, Y, I) = \{\langle A, A^{\uparrow}\rangle \mid A \in \text{Ext}(X, Y, I)\}$$
$$\mathscr{B}(X, Y, I) = \{\langle B^{\downarrow}, B\rangle \mid B \in \text{Int}(X, Y, I)\}$$

The above definition of Galois connection can be simplified by the following simplified form. $\langle f, g\rangle$ is a Galois connection between $X$ and $Y$ iff for every $A \subseteq X$ and $B \subseteq Y$:

$$A \subseteq g(B) \text{ iff } B \subseteq f(A).$$

Galois connections with respect to union and intersection satisfy the following properties: Let $\langle f, g\rangle$ be a Galois connection between $X$ and $Y$. For $A_j \subseteq X, j \in J$ and $B_j \subseteq Y, j \in J$, we have:

$$f(\bigcup_{j\in J} A_j) = \bigcap_{j\in J} f(A_j)$$
$$g(\bigcup_{j\in J} B_j) = \bigcap_{j\in J} g(B_j)$$

Every pair of concept-forming operators forms a Galois connection, and every Galois connection is a concept-forming operator of a particular formal context.

Let $\langle f, g\rangle$ be a Galois connection between $X$ and $Y$. Consider a formal context $\langle X, Y, I\rangle$ such that $I$ is defined by

$$\langle x, y\rangle \in I \text{ iff } y \in f(\{x\}) \text{ or equivalently, iff } x \in g(\{y\})$$

for each $x \in X$ and $y \in Y$. Then, $\langle {\uparrow_I}, {\downarrow_I}\rangle = \langle f, g\rangle$, i.e., $\langle {\uparrow_I}, {\downarrow_I}\rangle$ induced by $\langle X, Y, I\rangle$ coincide with $\langle f, g\rangle$.

We can establish representation result in the following form, i.e., $I \mapsto \langle {\uparrow_I}, {\downarrow_I}\rangle$ and $\langle {\uparrow_I}, {\downarrow_I}\rangle \mapsto I_{\langle\uparrow,\downarrow\rangle}$ are mutually inverse mappings between the set of all binary relations between $X$ and $Y$ and the set of all Galois connections between $X$ and $Y$.

We can also see the duality relationships between extents and intents. For $\langle A_1, B_1\rangle, \langle A_2, B_2\rangle \in \mathscr{B}(X, Y, I)$, we have that $A_1 \subseteq A_2$ iff $B_2 \subseteq B_1$. Then, we have the following peoperties:

(1) $\langle \text{Ext}(X, Y, I), \subseteq\rangle$ and $\langle \text{Int}(X, Y, I), \subseteq\rangle$ are partially ordered sets.
(2) $\langle \text{Ext}(X, Y, I), \subseteq\rangle$ and $\langle \text{Int}(X, Y, I), \subseteq\rangle$ are dually isomorphic, i.e., there is a mapping $f : \text{Ext}(X, Y, I) \to \text{Int}(X, Y, I)$ satisfying $A_1 \subseteq A_2$ iff $f(A_2) \subseteq f(A_1)$
(3) $\langle \mathscr{B}(X, Y, I), \leq\rangle$ is isomorphic to $\langle \text{Ext}(X, Y, I), \subseteq\rangle$
(4) $\langle \mathscr{B}(X, Y, I), \leq\rangle$ is dually isomorphic to $\langle \text{Int}(X, Y, I), \subseteq\rangle$.

We can also state the property of fixpoints of closure operators. For a closure operator $C$ on $X$, the partially ordered set $\langle \text{fix}(C), \subseteq\rangle$ of fixpoints of $C$ is a complete lattice with infima and suprema given by:

$$\bigwedge_{j\in J} A_j = C(\bigcap_{j\in J} A_j)$$
$$\bigvee_{j\in J} A_j = C(\bigcup_{j\in J} A_j)$$

The following is the main result of concept lattices due to Wille.

(1) $\mathscr{B}(X, Y, I)$ is a complete lattice with infima and suprema given by

$$\bigwedge_{j\in J} \langle A_j, B_j\rangle = \langle \bigcap_{j\in J} A_j, (\bigcup_{j\in J} B_j)^{\downarrow\uparrow}\rangle, \bigvee_{j\in J} \langle A_j, B_j\rangle = \langle (\bigcup_{j\in J} A_j)^{\uparrow\downarrow}, \bigcap_{j\in J} B_j\rangle.$$

(2) Moreover, an arbitrary complete lattice $\mathbf{V} = (V, \leq)$ is isomorphic to $\mathscr{B}(X, Y, I)$ iff there are mapping $\gamma : X \to V, \mu : Y \to X$ such that

(i) $\gamma(X)$ is $\bigvee$-dense in $V$, $\mu(Y)$ is $\bigwedge$-dense in $V$
(ii) $\gamma(x) \leq \mu(y)$ iff $\langle x, y\rangle \in I$

In formal concept analysis, we can clarify and reduce formal concepts by removing some of objects or attributes in a formal context. A formal context $\langle X, Y, I\rangle$ is called *clarified* if the corresponding table neither contain identical rows nor identical columns. Namely, if $\langle X, Y, I\rangle$ is clarified then:

$\{x_1\}^{\uparrow} = \{x_2\}^{\uparrow}$ implies $x_1 = x_2$ for every $x_1, x_2 \in X$,
$\{y_1\}^{\downarrow} = \{y_2\}^{\downarrow}$ implies $y_1 = y_2$ for every $y_1, y_2 \in Y$.

Clarification can be performed by removing identical rows and columns. If $\langle X_1, Y_1, I_1\rangle$ is a clarified context resulting from $\langle X_2, Y_2, I_2\rangle$ by clarification, then $\mathscr{B}(X_1, Y_1, I_1)$ is isomorphic to $\mathscr{B}(X_2, Y_2, I_2)$.

For a formal context $\langle X, Y, I\rangle$, an attribute $y \in Y$ is called *reducible* iff there is $Y' \subset Y$ with $y \notin Y'$ such that

$$\{y\}^{\downarrow} = \bigcap_{z \in Y'} \{z\}^{\downarrow}$$

i.e., the column corresponding to $y$ is the intersection of columns corresponding to $z$'s from $Y'$.

An object $x \in X$ is called *reducible* iff there is $X' \subset X$ with $x \notin X'$ such that

$$\{x\}^{\uparrow} = \bigcap_{z \in X'} \{z\}^{\uparrow}$$

i.e., the row corresponding to $x$ is the intersection of columns corresponding to $z$'s from $X'$.

Let $y \in Y$ be reducible in $\langle X, Y, I\rangle$. Then, $\mathscr{B}(X, Y - \{y\}, J)$ is isomorphic to $\mathscr{B}(X, Y, I)$, where $J = I \cap (X \times (Y - \{y\}))$ is the restriction of $I$ to $X \times Y - \{y\}$, i.e., $\langle X, Y - \{y\}, J\rangle$ results by removing $y$ from $\langle X, Y, I\rangle$.

$\langle X, Y, I\rangle$ is *row reducible* if no object $x \in X$ is reducible; it is *column reducible* if no attribute $y \in Y$ is reducible; it is *reduced* if it is both row reduced and column reduced.

*Arrow relations* can find which objects and attributes are reducible. For $\langle X, Y, I\rangle$, we define relations $\nearrow, \swarrow, \updownarrow$ between $X$ and $Y$:

$x \swarrow y$ iff $\langle x, y\rangle \notin I$ and if $\{x\}^{\uparrow} \subset \{x_1\}^{\uparrow}$ then $\langle x_1, y\rangle \in I$
$x \nearrow y$ iff $\langle x, y\rangle \notin I$ and if $\{x\}^{\downarrow} \subset \{x_1\}^{\downarrow}$ then $\langle x_1, y\rangle \in I$
$x \updownarrow y$ iff $x \swarrow y$ and $x \nearrow y$.

Thus, if $\langle x, y\rangle \in I$, then none of the above three relations occurs. Consequently, the arrow relations can be entered in the table of $\langle X, Y, I\rangle$. There is the following connections between arrow relations and reducibility.

$$\langle \{x\}^{\uparrow\downarrow}, \{x\}^{\uparrow}\rangle \text{ is } \bigvee\text{-irreducible iff there is } y \in Y \text{ such that } x \swarrow y$$

$$\langle \{y\}^{\downarrow}, \{y\}^{\downarrow\uparrow}\rangle \text{ is } \bigvee\text{-irreducible iff there is } x \in X \text{ such that } x \nearrow y.$$

Formal concept analysis can also deal with *attribute implication* concerning dependencies of data. Let $Y$ be a non-empty set of attributes.

An attribute implication over $Y$ is an expression

$$A \Rightarrow B$$

where $A, B \subseteq Y$.

An attribute implication $A \Rightarrow B$ over $Y$ is *true* (valid) in a set $M \subset Y$ iff $A \subseteq M$ implies $B \subseteq M$. We write $\|A \Rightarrow B\|_M = 1$ (0) if $A \Rightarrow B$ is true (false) in $M$.

Let $M$ be a set of attributes of some object $x$, $\|A \Rightarrow B\|_M = 1$ says "if $x$ has all attributes from $A$ then $x$ has all attributes from $B$", because "if $x$ has all attributes from $C$" is equivalent to $C \subseteq M$.

It is possible to extend the validity of $A \Rightarrow B$ to collections $\mathcal{M}$ of $M$'s (collections of subsets of attributes), i.e., define validity of $A \Rightarrow B$ in $\mathcal{M} \subseteq 2^Y$.

An attribute implication $A \Rightarrow B$ over $Y$ is true (valid) in $\mathcal{M}$ if $A \Rightarrow B$ is true in each $M \in \mathcal{M}$. An attribute implication $A \Rightarrow B$ over $Y$ is true (valid) in a table (formal context) $\langle X, Y, I\rangle$ iff $A \Rightarrow B$ is true in $\mathcal{M} = \{\{x\}^{\uparrow} \mid x \in X\}$.

We define semantic consequence (entailment). An attribute implication $A \Rightarrow B$ follows semantically from a theory $T$, denoted $T \models A \Rightarrow B$ iff $A \Rightarrow B$ is true in every model $M$ of $T$.

The system for reasoning about attribute implications consists of the following deduction rules:

(Ax) infer $A \cup B \Rightarrow A$,
(Cut) from $A \Rightarrow B$ and $B \cup C \Rightarrow D$ infer $A \cup C \Rightarrow D$.

Note that the above deduction rules are due to Armstrong's work on functional dependencies in databases; see Armstrong [14].

A *proof* of $A \Rightarrow B$ from a set $T$ of attribute implications is a sequence $A_1 \Rightarrow B_1, \ldots, A_n \Rightarrow B_n$ of attribute implications satisfying:

(1) $A_n \Rightarrow B_n$ is just $A \Rightarrow B$,
(2) for every $i = 1, 2, \ldots, n$:

either $A_i \Rightarrow B_i \in T$ (assumption)
or $A_i \Rightarrow B_i$ results by using (Ax) or (Cut) from preceding attribute implications $A_j \Rightarrow B_j$'s (deduction)

If we have a proof of $A \Rightarrow B$ from $T$, then we write $T \vdash A \Rightarrow B$. We have the following derivable rules:

(Ref) infer $A \Rightarrow A$,
(Wea) from $A \Rightarrow B$, infer $A \cup C \Rightarrow B$,
(Add) from $A \Rightarrow B$ and $A \Rightarrow C$, infer $A \Rightarrow B \cup C$,
(Pro) from $A \Rightarrow B \cup C$, infer $A \Rightarrow B$,
(Tra) from $A \Rightarrow B$ and $B \Rightarrow C$, infer $A \Rightarrow C$,

for every $A, B, C, D \subseteq Y$.

We can show that (Ax) and (Cut) are sound. It is also possible to prove soundness of above derived rules.

We can define two notions of consequence, i.e., *semantic consequence* and *syntactic consequence*:

Semantic: $T \models A \Rightarrow B (A \Rightarrow B$ semantically follows from $T)$
Syntactic: $T \vdash A \Rightarrow B (A \Rightarrow B$ syntactically follows from $T)$

*Semantic closure* of $T$ is the set

$$sem(T) = \{A \Rightarrow B \mid T \models A \Rightarrow B\}$$

of all attribute implications which semantically follows from $T$.

*Syntactic closure* of $T$ is the set

$$syn(T) = \{A \Rightarrow B \mid T \vdash A \Rightarrow B\}$$

of all attribute implications which syntactically follows from $T$.

$T$ is *semantically closed* if $T = sem(T)$. $T$ is *syntactically closed* if $T = syn(T)$. Note that $sem(T)$ is the least set of attribute implications which is semantically closed containinig $T$ and that $syn(T)$ is the least set of attribute implications which is syntactically closed containinig $T$.

It can be proved that $T$ is syntactically closed iff for any $A, B, C, D \subseteq Y$

(1) $A \cup B \Rightarrow B \in T$,
(2) if $A \Rightarrow B \in T$ and $B \cup C \Rightarrow D \in T$ implies $A \cup C \Rightarrow D \in T$.

Then, if $T$ is semantically closed, then $T$ is syntactically closed. It can also be proved that if $T$ is syntactically closed, then $T$ is semantically closed. Consequently, soundness and completeness follow:

$$T \vdash A \Rightarrow B \text{ iff } T \models A \Rightarrow B.$$

We turn to models of attribute implications. For a set $T$ of attribute implications, denote

$$Mod(T) = \{M \subseteq Y \mid \|A \Rightarrow B\|_M = 1 \ for \ every \ A \Rightarrow B \in T\}$$

That is, $Mod(T)$ is the set of all models of $T$.

A closure system in a set of $Y$ is any system $\mathscr{S}$ of subsets of $Y$ which contains $Y$ and is closed under arbitrary intersections. That is, $Y \in \mathscr{S}$ and $\bigcap \mathscr{R} \in \mathscr{S}$ for every $\mathscr{R} \subseteq \mathscr{S}$ (intersection of every subsystem $\mathscr{R}$ of $\mathscr{S}$ belongs to $\mathscr{S}$.

There is a one-to-one relationship between closure systems in $Y$ and closure operators in $Y$. Namely, for a closure operator $C$ in $Y$, $\mathscr{S}_C = \{A \in 2^X \mid A = C(A)\} = \text{fix}(C)$ is a closure system in $Y$.

Given a closure system in $Y$, we set

$$C_{\mathscr{S}}(A) = \bigcap \{B \in \mathscr{S} \mid A \subseteq B\}$$

for any $A \subseteq X$, $C_{\mathscr{S}}$ is a closure operator on $Y$. This is a one-to-one relationship, i.e., $C = C_{\mathscr{S}_C}$ and $\mathscr{S} = \mathscr{S}_{C_{\mathscr{S}}}$.

It is shown that for a set of $T$ of attribute implications, $Mod(T)$ is a closure system in $Y$. Since Mod(T) is a closure system, we can consider the corresponding closure operator $C_{\text{Mod}(T)}$, i.e., the fixpoints of $C_{\text{Mod}(T)}$ are just models of $T$.

Therefore, for every $A \subseteq Y$, there exists the least model of $Mod(T)$ which contains $A$, namely such least models is just $C_{\text{Mod}(T)}(A)$.

We can test entailment via least models follows. For any $A \Rightarrow B$ and any $T$, we have:

$$T \models A \Rightarrow B \text{ iff } \|A \Rightarrow B\|_{C_{Mod(T)}(A)} = 1.$$

It follows that the deductive system for attribute implications is sound and complete. And it can serve as a basis for reasoning about dependencies.

As reviewed in this section, formal concept analysis offers an interesting tool for data analysis. It has a mathematical foundation based on concept lattice with reasoning mechanisms based on attribute implications. In addition, formal concept analysis can visualize data.

Because formal concept analysis uses the notion of table, some similarities with rough set theory may be found. In fact, it uses classical (two-valued) basis. However, it is not clear whether it is possible to modify it, relating to some non-classical logic.

## 2.5 Decision Logic

Pawlak developed *decision logic* ($DL$) for reasoning about knowledge. His main goal is reasoning about knowledge concerning reality. Knowledge is represented as a value-attribute table, called *knowledge representation system*.

There are several advantages to represent knowledge in tabular form. The data table can be interpreted differently, namely it can be formalized as a logical system. The idea leads to decision logic.

The language of $DL$ consists of atomic formulas, which are attribute-value pairs, combined by logical connectives to form compound formulas. The alphabet of the language consists of:

1. $A$: the set of attribute constants
2. $V = \bigcup V_a$: the set of attribute constants $a \in A$
3. Set $\{\sim, \vee, \wedge, \rightarrow, \equiv\}$ of propositional connectives, called negation, disjunction, conjunction, implication and equivalence, respectively.

The set of formulas in $DL$-language is the least set satisfying the following conditions:

1. Expressions of the form $(a, v)$, or in short $a_v$, called atomic formulas, are formulas of DL-language for any $a \in A$ and $v \in V_a$.
2. If $\phi$ and $\psi$ are formulas of $DL$-language, then so are $\sim \phi$, $(\phi \vee \psi)$, $(\phi \wedge \psi)$, $(\phi \rightarrow \psi)$ and $(\phi \equiv \psi)$.

Formulas are used as descriptions of objects of the universe. In particular, atomic formula of the form $(a, v)$ is interpreted as a description of all objects having value $v$ for attribute $a$.

The semantics for $DL$ is given by a model. For $DL$, the model is KR-system $S = (U, A)$, which describes the meaning of symbols of predicates $(a, v)$ in $U$, and if we properly interpret formulas in the model, then each formula becomes a meaningful sentence, expressing properties of some objects.

An object $x \in U$ *satisfies* a formula $\phi$ in $S = (U, A)$, denoted $x \models_S \phi$ or in short $x \models \phi$, iff the following conditions are satisfied:

1. $x \models (a, v)$ iff $a(x) = v$
2. $x \models \sim \phi$ iff $x \not\models \phi$
3. $x \models \phi \vee \psi$ iff $x \models \phi$ or $x \models \psi$
4. $x \models \phi \wedge \psi$ iff $x \models \phi$ and $x \models \psi$

The following are clear from the above truth definition:

5. $x \models \phi \rightarrow \psi$ iff $x \models \sim \phi \vee \psi$
6. $x \models \phi \equiv \psi$ iff $x \models \phi \rightarrow \psi$ and $x \models \psi \rightarrow \phi$

If $\phi$ is a formula, then the set $|\phi|_S$ defined as follows:

$$|\phi|_s = \{x \in U \mid x \models_S \phi\}$$

will be called the *meaning* of the formula $\phi$ in $S$.

**Proposition 2.19** *The meaning of arbitrary formulas satisfies the following:*

$$|(a, v)|_S = \{x \in U \mid a(x) = v\}$$
$$|\sim \phi|_S = -|\phi|_S$$
$$|\phi \vee \psi|_S = |\phi|_S \cup |\psi|_S$$
$$|\phi \wedge \psi|_S = |\phi|_S \cap |\psi|_S$$
$$|\phi \rightarrow \psi|_S = -|\phi|_S \cup |\psi|_S$$

$|\phi \equiv \psi|_S = (|\phi|_S \cap |\psi|_S) \cup (-|\phi|_S \cap -|\psi|_S)$

Thus, the meaning of the formula $\phi$ is the set of all objects having the property expressed by the formula $\phi$, or the meaning of the formula $\phi$ is the description in the KR-language of the set objects $|\phi|$.

A formula $\phi$ is said to be *true* in a KR-system $S$, denoted $\models_S \phi$, iff $|\phi|_S = U$, i.e., the formula is satisfied by all objects of the universe in the system $S$. Formulas $\phi$ and $\psi$ are equivalent in $S$ iff $|\phi|_S = |\psi|_S$.

**Proposition 2.20** *The following are the simple properties of the meaning of a formula.*

$\models_S \phi$ *iff* $|\phi| = U$
$\models_S \sim \phi$ *iff* $|\phi| = \emptyset$
$\phi \rightarrow \psi$ *iff* $|\psi| \subseteq |\psi|$
$\phi \equiv \psi$ *iff* $|\psi| = |\psi|$

The meaning of the formula depends on the knowledge we have about the universe, i.e., on the knowledge representation system. In particular, a formula may be true in one knowledge representation system, but false in another one.

However, there are formulas which are true independent of the actual values of attributes appearing them. But, they depend only on their formal structure.

Note that in order to find the meaning of such a formula, one need not be acquainted with the knowledge contained in any specific knowledge representation system because their meaning is determined by its formal structure only.

Hence, if we ask whether a certain fact is true in light of our actual knowledge, it is sufficient to use this knowledge in an appropriate way. For formulas which are true (or not) in every possible knowledge representation system, we do not need in any particular knowledge, but only suitable logical tools.

To deal with deduction in $DL$, we need suitable axioms and inference rules. Here, axioms will correspond closely to axioms of classical propositional logic, but some specific axioms for the specific properties of knowledge representation systems are also needed. The only inference rule will be *modus ponens*.

We will use the following abbreviations:

$$\phi \wedge \sim \phi =_{\text{def}} 0$$

$$\phi \vee \sim \phi =_{\text{def}} 1$$

Obviously, $\models 1$ and $\models \sim 0$. Thus, 0 and 1 can be assumed to denote *falsity* and *truth*, respectively.

Formula of the form:

$$(a_1, v_1) \wedge (a_2, v_2) \wedge \ldots \wedge (a_n, v_n)$$

where $v_{a_i} \in V_a$, $P = \{a_1, a_2, \ldots, a_n\}$ and $P \subseteq A$ is called a *P-basic formula* or in short $P$-formula. Atomic formulas is called *A-basic formula* or in short basic formula.

Let $P \subseteq A, \phi$ be a $P$-formula and $x \in U$. If $x \models \phi$ then $\phi$ is called the *P-description* of $x$ in $S$. The set of all $A$-basic formulas satisfiable in the knowledge representation system $S = (U, A)$ is called the *basic knowledge* in $S$.

We write $\sum_S(P)$, or in short $\sum(P)$, to denote the disjunction of all $P$-formulas satisfied in $S$. If $P = A$ then $\sum(A)$ is called the *characteristic formula* of $S$.

The knowledge representation system can be represented by a data table. And its columns are labelled by attributes and its rows are labelled by objects. Thus, each row in the table is represented by a certain $A$-basic formula, and the whole table is represented by the set of all such formulas. In $DL$, instead of tables, we can use sentences to represent knowledge.

There are specific axioms of $DL$:

1. $(a, v) \wedge (a, u) \equiv 0$ for any $a \in A, u, v \in V$ and $v \neq u$
2. $\bigvee_{v \in V_a} (a, v) \equiv 1$ for every $a \in A$
3. $\sim (a, v) \equiv \bigvee_{a \in V_a, u \neq v} (a, u)$ for every $a \in A$

The axiom (1) states that each object can have exactly one value of each attribute.

The axiom (2) assumes that each attribute must take one of the values of its domain for every object in the system.

The axiom (3) allows us to eliminate negation in such a way that instead of saying that an object does not possess a given property we can say that it has one of the remaining properties.

**Proposition 2.21** *The following holds for $DL$:*

$$\models_S \sum_S(P) \equiv 1 \textit{ for any } P \subseteq A.$$

Proposition 2.21 means that the knowledge contained in the knowledge representation system is the whole knowledge available at the present stage. and corresponds to the so-called *closed world assumption* (CWA).

We say that a formula $\phi$ is *derivable* from a set of formulas $\Omega$, denoted $\Omega \vdash \phi$, iff it is derivable from axioms and formulas of $\Omega$ by finite application of *modus ponens*. Formula $\phi$ is a theorem of $DL$, denoted $\vdash \phi$, if it is derivable from the axioms only. A set of formulas $\Omega$ is *consistent* iff the formula $\phi \wedge \sim \phi$ is not derivable from $\Omega$.

Note that the set of theorems of $DL$ is identical with the set of theorems of classical propositional logic with specific axioms (1)–(3), in which negation can be eliminated.

Formulas in the *KR*-language can be represented in a special form called *normal form*, which is similar to that in classical propositional logic.

Let $P \subseteq A$ be a subset of attributes and let $\phi$ be a formula in *KR*-language. We say that $\phi$ is in a *P-normal form* in $S$, in short in $P$-normal form, iff either $\phi$ is 0 or $\phi$ is 1, or $\phi$ is a disjunction of non-empty $P$-basic formulas in $S$. (The formula $\phi$ is non-empty if $|\phi| \neq \emptyset$).

**Table 2.1** *KR*-system 1

| *U* | *a* | *b* | *c* |
|---|---|---|---|
| 1 | 1 | 0 | 2 |
| 2 | 2 | 0 | 3 |
| 3 | 1 | 1 | 1 |
| 4 | 1 | 1 | 1 |
| 5 | 2 | 1 | 3 |
| 6 | 1 | 0 | 3 |

$A$-normal form will be referred to as *normal form*. The following is an important property in the $DL$-language.

**Proposition 2.22** *Let $\phi$ be a formula in $DL$-language and let $P$ contain all attributes occurring in $\phi$. Moreover, (1)–(3) and the formula $\sum_S(A)$. Then, there is a formula $\psi$ in the $P$-normal form such that $\phi \equiv \psi$.*

Here is the example from Pawlak [1]. Consider the following *KR*-system (Table 2.1).

The following $a_1b_0c_2$, $a_2b_0c_3$, $a_1b_1c_1$, $a_2b_1c_3$, $a_1b_0c_3$ are all basic formulas (basic knowledge) in the *KR*-system. For simplicity, we will omit the symbol of conjunction $\wedge$ in basic formulas.

The characteristic formula of the system is:

$$a_1b_0c_2 \vee a_2b_0c_3 \vee a_1b_1c_1 \vee a_2b_1c_3 \vee a_1b_0c_3$$

Here, we give the following meanings of some formulas in the system:

$$|a_1 \vee b_0c_2| = \{1, 3, 4, 6\}$$
$$|\sim (a_2b_1)| = \{1, 2, 3, 4, 6\}$$
$$|b_0 \rightarrow c_2| = \{1, 3, 4, 5\}$$
$$|a_2 \equiv b_0| = \{2, 3, 4\}$$

Below are given normal forms of formulas considerred in the above example for *KR*-system 1:

$$a_1 \vee b_0c_2 = a_1b_0c_2 \vee a_1b_1c_1 \vee a_1b_0c_3$$
$$\sim (a_2b_1) = a_1b_0c_2 \vee a_2b_0c_3 \vee a_1b_1c_1 \vee a_1b_0c_3$$
$$b_0 \rightarrow c_2 = a_1b_0c_2 \vee a_1b_1c_1 \vee a_2b_1c_3$$
$$a_2 \equiv b_0 = a_2b_0c_1 \vee a_2b_0c_2 \vee a_2b_0c_3 \vee a_1b_1c_1 \vee a_1b_1c_2 \vee a_1b_1c_3$$

Examples of formulas in $\{a, b\}$-normal form are:

$$\sim (a_2b_1) = a_1b_0 \vee a_2b_0 \vee a_1b_1 \vee a_1b_0$$
$$a_2 \equiv b_0 = a_2b_0 \vee a_1b_1$$

The following is an example of a formula in $\{b, c\}$-normal form:

$$b_0 \rightarrow c_2 = b_0c_2 \vee b_1c_1 \vee b_1c_3$$

Thus, in order to compute the normal form of a formula, we have to transform by using propositional logic and the specific axioms for a given *KR*-system.

Any implication $\phi \rightarrow \psi$ is called a *decision rule* in the *KR*-language. $\phi$ and $\psi$ are referred to as the *predecessor* and *successor* of $\phi \rightarrow \psi$, respectively.

If a decision rule $\phi \rightarrow \psi$ is true in $S$, we say that the decision rule is *consistent* in $S$; otherwise the decision rule is *inconsistent* in $S$.

If $\phi \rightarrow \psi$ is a decision rule and $\phi$ and $\psi$ are $P$-basic and $Q$-basic formulas respectively, then the decision rule $\phi \rightarrow \psi$ is called a *PQ-basic decision rule* (in short *PQ*-rule).

A *PQ*-rule $\phi \rightarrow \psi$ is *admissible* in $S$ if $\phi \wedge \psi$ is satisfiable in $S$.

**Proposition 2.23** *A PQ-rule is true (consistent) in S iff all $\{P, Q\}$-basic formulas which occurr in the $\{P, Q\}$-normal form of the predecessor of the rule, also occurr in $\{P, Q\}$-normal form of the successor of the rule; otherwise the rule is false (inconsistent).*

The rule $b_0 \rightarrow c_2$ is false in the above table for *KR*-system 1, since the $\{b, c\}$-normal form of $b_0$ is $b_0c_2 \vee b_0c_3$, $\{b, c\}$-normal form of $c_2$ is $b_0c_2$, and the formula $b_0c_3$ does not occur in the successor of the rule.

On the other hand, the rule $a_2 \rightarrow c_3$ is true in the table, because the $\{a, c\}$-normal form of $a_2$ is $a_2c_3$, whereas the $\{a, c\}$-normal form of $c_3$ is $a_2c_3 \vee a_1c_3$.

Any finite set of decision rules in a $DL$-language is referred to as a *decision algorithm* in the $DL$-language. If all decision rules in a basic decision algorithm are *PQ*-decision rules, then the algorithm is said to be *PQ-decision algorithm*, or in short *PQ*-algorithm, and will be denoted by $(P, Q)$.

A *PQ*-algorithm is *admissible* in $S$, if the algorithm is the set of all *PQ*-rules admissible in $S$.

A *PQ*-algorithm is *complete* in $S$, iff for every $x \in U$ there exists a *PQ*-decision rule $\phi \rightarrow \psi$ in the algorithm such that $x \models \phi \wedge \psi$ in $S$; otherwise the algorithm is *incomplete* in $S$.

A *PQ*-algorithm is *consistent* in $S$ iff all its decision rules are consistent (true) in $S$; otherwise the algorithm is *inconsistent*.

Sometimes consistency (inconsistency) may be interpreted as *determinism* (*indeterminism*).

Given a *KR*-system, any two arbitrary, non-empty subset of attributes $P$, $Q$ in the system determines uniquely a *PQ*-decision algorithm.

Consider the following *KR*-system from Pawlak [1].

Assume that $P = \{a, b, c\}$ and $Q = \{d, e\}$ are condition and decision attributes, respectively. Set $P$ and $Q$ uniquely associate the following *PQ*-decision algorithm with the table.

**Table 2.2** *KR*-system 2

| *U* | *a* | *b* | *c* | *d* | *e* |
|---|---|---|---|---|---|
| 1 | 1 | 0 | 2 | 1 | 1 |
| 2 | 2 | 1 | 0 | 1 | 0 |
| 3 | 2 | 1 | 2 | 0 | 2 |
| 4 | 1 | 2 | 2 | 1 | 1 |
| 5 | 1 | 2 | 0 | 0 | 2 |

$a_1b_0c_2 \rightarrow d_1e_1$
$a_2b_1c_0 \rightarrow d_1e_0$
$a_2b_1c_2 \rightarrow d_0e_2$
$a_1b_2c_2 \rightarrow d_1e_1$
$a_1b_2c_0 \rightarrow d_0e_2$

If assume that $R = \{a, b\}$ and $T = \{c, d\}$ are condition and decision attributes, respectively, then the $RT$-algorithm determined by Table 2.2 is the following:

$a_1b_0 \rightarrow c_2d_1$
$a_2b_1 \rightarrow c_0d_1$
$a_2b_1 \rightarrow c_2d_0$
$a_1b_2 \rightarrow c_2d_1$
$a_1b_2 \rightarrow c_0d_0$

Of course, both algorithms are admissible and complete.

In order to check whether or not a decision algorithm is consistent, we have to check whether all its decision rules are true. The following proposition gives a much simpler method to solve this problem.

**Proposition 2.24** *A PQ-decision rule* $\phi \rightarrow \psi$ *in a PQ-decision algorithm is consistent (true) in S iff for any PQ-decision rule* $\phi' \rightarrow \psi'$ *in PQ-decision algorithm,* $\phi = \phi'$ *implies* $\psi = \psi'$.

In Proposition 2.24, order of terms is important, since we require equality of expressions. Note also that in order to check whether or not a decision rule $\phi \rightarrow \psi$ is true we have to show that the predecessor of the rule (the formula $\phi$) discerns the decision class $\psi$ from the remaining decision classes of the decision algorithm in question. Thus, the concept of truth is somehow replaced by the concept of indiscernibility.

Consider the *KR*-system 2 again. With $P = \{a, b, c\}$ and $Q = \{d, e\}$ as condition and decision attributes. Let us check whether the *PQ*-algorithm:

**Table 2.3** *KR*-system 2

| $U$ | $a$ | $b$ | $c$ | $d$ | $e$ |
|---|---|---|---|---|---|
| 1 | 1 | 0 | 2 | 1 | 1 |
| 4 | 1 | 2 | 2 | 1 | 1 |
| 2 | 2 | 1 | 0 | 1 | 0 |
| 3 | 2 | 1 | 2 | 0 | 2 |
| 5 | 1 | 2 | 0 | 0 | 2 |

$$a_1 b_0 c_2 \rightarrow d_1 e_1$$
$$a_2 b_1 c_0 \rightarrow d_1 e_0$$
$$a_2 b_1 c_2 \rightarrow d_0 e_2$$
$$a_1 b_2 c_2 \rightarrow d_1 e_1$$
$$a_1 b_2 c_0 \rightarrow d_0 e_2$$

is consistent or not. Because the predecessors of all decision rules in the algorithm are different (i.e., all decision rules are discernible by predecessors of all decision rules in the algorithm), all decision rules in the algorithm are consistent (true) and cosequently the algorithm is consistent.

This can be also seen directly from Table 2.3.

The $RT$-algorithm, where $R = \{a, b\}$ and $T\{c, d\}$

$$a_1 b_0 \rightarrow c_2 d_1$$
$$a_2 b_1 \rightarrow c_0 d_1$$
$$a_2 b_1 \rightarrow c_2 d_0$$
$$a_1 b_2 \rightarrow c_2 d_1$$
$$a_1 b_2 \rightarrow c_0 d_0$$

is inconsistent bacause the rules

$$a_2 b_1 \rightarrow c_0 d_1$$
$$a_2 b_1 \rightarrow c_2 d_0$$

have the same predecessors and different successors, i.e., we are unable to discern $c_0 d_1$ and $c_2 d_0$ by means of condition $a_2 b_1$. Thus, both rules are inconsistent (false) in the *KR*-system. Similarly, the rules

$$a_1 b_2 \rightarrow c_2 d_1$$
$$a_1 b_2 \rightarrow c_0 d_0$$

are also inconsistent (false).

We turn to *dependency* of attributes. Formally, the dependency is defined as below. Let $K = (U, \mathbf{R})$ be a knowledge base and $\mathbf{P}, \mathbf{Q} \subseteq \mathbf{R}$.

(1) Knowledge $\mathbf{Q}$ *depends on knowledge* $\mathbf{P}$ iff $IND(\mathbf{P}) \subseteq IND(\mathbf{Q})$.
(2) Knowlledge $\mathbf{P}$ and $\mathbf{Q}$ are *equivalent*, denoted $\mathbf{P} \equiv \mathbf{Q}$, if $\mathbf{P} \Rightarrow \mathbf{Q}$ and $\mathbf{Q} \Rightarrow \mathbf{P}$.
(3) Knowledge $\mathbf{P}$ and $\mathbf{Q}$ are *independent*, denoted $\mathbf{P} \not\equiv \mathbf{Q}$, iff neither $\mathbf{P} \Rightarrow \mathbf{Q}$ nor $\mathbf{Q} \Rightarrow \mathbf{P}$ hold.

Obviously, $\mathbf{P} \equiv \mathbf{Q}$ iff $IND(\mathbf{P}) \equiv IND(\mathbf{Q})$.

The dependency can be interpreted in different ways as Proposition 2.25 indicates:

**Proposition 2.25** *The following conditions are equivalent:*

*(1)* $\mathbf{P} \Rightarrow \mathbf{Q}$
*(2)* $IND(\mathbf{P} \cup \mathbf{Q}) = INS(\mathbf{P})$
*(3)* $POS_{\mathbf{P}}(\mathbf{Q}) = U$
*(4)* $\underline{\mathbf{P}}X$ *for all* $X \in U/\mathbf{Q}$

*where* $\underline{\mathbf{P}}X$ *denotes* $\underline{IND(\mathbf{P})/X}$.

By Proposition 2.25, we can see the following: if $\mathbf{Q}$ depends on $\mathbf{P}$ then knowledge $\mathbf{Q}$ is superflous within the knowledge base in the sense that the knowledge $\mathbf{P} \cup \mathbf{Q}$ and $\mathbf{P}$ provide the same characterization of objects.

**Proposition 2.26** *If* $\mathbf{P}$ *is a reduct of* $\mathbf{Q}$*, then* $\mathbf{P} \Rightarrow \mathbf{Q} - \mathbf{P}$ *and* $IND(\mathbf{P}) = IND(\mathbf{Q})$.

**Proposition 2.27** *The following hold.*

*(1) If* $\mathbf{P}$ *is dependent, then there exists a subset* $\mathbf{Q} \subset \mathbf{P}$ *such that* $\mathbf{Q}$ *is a reduct of* $\mathbf{P}$.
*(2) If* $\mathbf{P} \subseteq \mathbf{Q}$ *and* $\mathbf{P}$ *is independent, then all basic relations in* $\mathbf{P}$ *are pairwise independent.*
*(3) If* $\mathbf{P} \subseteq \mathbf{Q}$ *and* $\mathbf{P}$ *is independent, then every subset* $\mathbf{R}$ *of* $\mathbf{P}$ *is independent.*

**Proposition 2.28** *The following hold:*

*(1) If* $\mathbf{P} \Rightarrow \mathbf{Q}$ *and* $\mathbf{P}' \supset \mathbf{P}$*, then* $\mathbf{P}' \Rightarrow \mathbf{Q}$.
*(2) If* $\mathbf{P} \Rightarrow \mathbf{Q}$ *and* $\mathbf{Q}' \subset \mathbf{Q}$*, then* $\mathbf{P} \Rightarrow \mathbf{Q}'$.
*(3)* $\mathbf{P} \Rightarrow \mathbf{Q}$ *and* $\mathbf{Q} \Rightarrow \mathbf{R}$ *imply* $\mathbf{P} \Rightarrow \mathbf{R}$.
*(4)* $\mathbf{P} \Rightarrow \mathbf{R}$ *and* $\mathbf{Q} \Rightarrow \mathbf{R}$ *imply* $\mathbf{P} \cup \mathbf{Q} \Rightarrow \mathbf{R}$.
*(5)* $\mathbf{P} \Rightarrow \mathbf{R} \cup \mathbf{Q}$ *imply* $\mathbf{P} \Rightarrow \mathbf{R}$ *and* $\mathbf{P} \cup \mathbf{Q} \Rightarrow \mathbf{R}$.
*(6)* $\mathbf{P} \Rightarrow \mathbf{Q}$ *and* $\mathbf{Q} \cup \mathbf{R} \Rightarrow \mathbf{T}$ *imply* $\mathbf{P} \cup \mathbf{R} \Rightarrow \mathbf{T}$
*(7)* $\mathbf{P} \Rightarrow \mathbf{Q}$ *and* $\mathbf{R} \Rightarrow \mathbf{T}$ *imply* $\mathbf{P} \cup \mathbf{R} \Rightarrow \mathbf{Q} \cup \mathbf{T}$.

The derivation (dependency) can be partial, which means that only part of knowledge $\mathbf{Q}$ is derivable from knowledge $\mathbf{P}$. We can define the partial derivability using the notion of the positive region of knowledge.

Let $K = (U, \mathbf{R})$ be the knowledge base and $\mathbf{P}, \mathbf{Q} \subset \mathbf{R}$. Knowledge $\mathbf{Q}$ depends in a degree $k$ $(0 \leq k \leq 1)$ from knowledge $\mathbf{P}$, in symbol $\mathbf{P} \Rightarrow_k \mathbf{Q}$, iff

$$k = \gamma_{\mathbf{P}}(\mathbf{Q}) = \frac{card(POS_{\mathbf{P}}(\mathbf{Q}))}{card(U)}$$

where *card* denotes cardinality of the set.

If $k = 1$, we say that **Q** *totally depends from* **P**; if $0 < k < 1$, we say that **Q** *roughly (partially) depends from* **P**, and if $k = 1$ we say that **Q** is *totally independent from* **P**, If $\mathbf{P} \Rightarrow_1 \mathbf{Q}$, we shall also write $\mathbf{P} \Rightarrow \mathbf{Q}$.

The above ideas can also be interpreted as an ability to classify objects. More precisely, if $k = 1$, then all elements of the universe can be classified to elementary categories of $U/\mathbf{Q}$ by using knowledge **P**.

Thus, the coefficient $\gamma_{\mathbf{P}}(\mathbf{Q})$ can be understood as a degree of dependency between, **Q** and **P**. In other words, if we restrict the set of objects in the knowledge base to the set $POS_{\mathbf{P}}(\mathbf{Q})$, we would obtain the knowledge base in which $\mathbf{P} \Rightarrow \mathbf{Q}$ is a total dependency.

The measure $k$ of dependency $\mathbf{P} \Rightarrow_k \mathbf{Q}$ does not capture how this partial dependency is actually distributed among classes of $U/\mathbf{Q}$. For example, some decision classes can be fully characterized by **P**, whereas others may be characteriaed only partially.

We will also need a coefficient $\gamma(X) = card(\underline{\mathbf{P}}X)/card(X)$ where $X \in U/\mathbf{Q}$ which shows how many elements of each class of $U/\mathbf{Q}$ can be classified by emplying knowledge **P**.

Thus, the two numbers $\gamma(\mathbf{Q})$ and $\gamma(X)$, $X \in U/\mathbf{Q}$ give us full information about "classification power" of the knowledge **P** with respect to the classification $U/\mathbf{Q}$.

**Proposition 2.29** *The following hold:*

*(1) If* $\mathbf{R} \Rightarrow_k \mathbf{P}$ *and* $\mathbf{Q} \Rightarrow_l \mathbf{P}$, *then* $\mathbf{R} \cup \mathbf{Q} \Rightarrow_m \mathbf{P}$, *for some* $m \geq max(k, l)$.
*(2) If* $\mathbf{R} \cup \mathbf{P} \Rightarrow_k \mathbf{Q}$ *then* $\mathbf{R} \Rightarrow_l \mathbf{Q}$ *and* $\mathbf{P} \Rightarrow_m \mathbf{Q}$, *for some* $l, m, \leq k$.
*(3) If* $\mathbf{R} \Rightarrow_k \mathbf{Q}$ *and* $\mathbf{R} \Rightarrow_l \mathbf{P}$, *then* $\mathbf{R} \Rightarrow_m \mathbf{Q} \cup \mathbf{P}$, *for some* $m \leq max(k, l)$.
*(4) If* $\mathbf{R} \Rightarrow_k \mathbf{Q} \cup \mathbf{P}$, *then* $\mathbf{R} \Rightarrow_l \mathbf{Q}$ *and* $\mathbf{R} \Rightarrow_m \mathbf{P}$, *for some* $l, m \geq k$.
*(5) If* $\mathbf{R} \Rightarrow_k \mathbf{P}$ *and* $\mathbf{P} \Rightarrow_l \mathbf{Q}$, *then* $\mathbf{R} \Rightarrow_m \mathbf{Q}$, *for some* $m \geq l + k - 1$.

Here, we return to the decision algorithm for dependency. We say that the set of attributes $Q$ depends *totally*, (or in short depends) on the set of attributes $P$ in $S$, if there exists a consistent $PQ$-algorithm in $S$. If $Q$ depends on $P$ in $S$, we write $P \Rightarrow_S S$, or in short $P \Rightarrow Q$.

We can also define partial dependency of attributes. We say that the set of attributes $Q$ *depends partially* on the set of attributes $P$ in $S$, if there exists an inconsistent $PQ$-algorithm in $S$.

The degree of *dependency* between attributes can be defind. Let $(P, Q)$ be a $PQ$-algorithm in $S$. By a *positive region* of the algorithm $(P, Q)$, denoted $POS(P, Q)$, we mean the set of all consistent (true) $PQ$-rules in the algorithm.

The positive region of the decision algorithm $(P, Q)$ is the consistent part (possibly empty) of the inconsistent algorithm. Obviously, a $PQ$-algorithm is inconsistent iff $POS(P, Q) \neq (P, Q)$ or what is the same $card(POS(P, Q)) \neq card(P, Q)$.

With every $PQ$-decision algorithm, we can associate a number
$k = card(POS(P, Q))/card(P, Q)$, called the degree of consistency, of the algorithm, or in short the degree of the algorithm, and we say that the $PQ$-algorithm has the degree (of consistency) $k$.

Obviously, $0 \leq k \leq 1$. If a $PQ$-algorithm has degree $k$, we can say that the set of attributes $Q$ *depend in degree k* on the set of attributes $P$, denoted $P \Rightarrow_k Q$.

Naturally, the algorithm is consistent iff $k = 1$; otherwise, i.e., if $k \neq 1$, the algorithm. All these concepts are the same as in those discussed above. Note that in the consistent algorithm all decisions are uniquely determined by conditions in the decision algorithm. In other words, this means that all decisions in a consistent algorithm are discernible by means of conditions available in the decision algorithm.

Decision logic provides a simple means for reasoning about knowledge only by using propositional logic, and is suitable to some applications. Note here that the so-called *decision table* can serve as a *KR*-system.

However, the usuability of decision logic seems to be restrictive. In other words, it is far from a general system for reasoning in general. In this book, we will lay general frameworks for reasoning based on rough set theory.

## 2.6 Reduction of Knowledge

One of the important problems in rough set theory is whether the whole knowledge is always necessary to define some categories available in the knowledge considered. This problem is called *knowledge reduction.*

There are two basic concepts in reduction of knowledge, i.e., *reduct* and *core*. Intuitively, a reduct of knowledge is its essential part, which is sufficient to define all basic concepts in the considered knowledge. The core is the set of the most characteristic part of knowledge.

Let $\mathbf{R}$ be a family of equivalence relations and let $R \in \mathbf{R}$. We say that $R$ is *dispensable* in $\mathbf{R}$ if $IND(\mathbf{R}) = IND(\mathbf{R} - \{R\})$; otherwise is *indispensable* in $\mathbf{R}$. The family $\mathbf{R}$ is *independent* if each $R \in \mathbf{R}$ is indispensable in $\mathbf{R}$; otherwise $\mathbf{R}$ is *dependent.*

**Proposition 2.30** *If $\mathbf{R}$ is independent and $\mathbf{P} \subseteq \mathbf{R}$, then $\mathbf{P}$ is also independent.*

The following proposition states the relationship between the core and reducts.

$\mathbf{Q} \subseteq \mathbf{P}$ is a *reduct* of $\mathbf{P}$ if $\mathbf{Q}$ is indenpendent and $IND(\mathbf{Q}) = IND(\mathbf{P})$. Obviously, $\mathbf{P}$ may have many reducts. The set of all indispensable relations in $\mathbf{P}$ is called the *core* of $\mathbf{P}$ denoted $CORE(\mathbf{P})$.

**Proposition 2.31** *$CORE(\mathbf{P}) = \bigcap RED(\mathbf{P})$, where $RED(\mathbf{P})$ is the famiy of all reducts of $\mathbf{P}$.*

Here is an example from Pawlak [1]. Suppose $\mathbf{R} = \{P, Q, R\}$ of three equivalence relations $P$, $Q$ and $R$ with the following equivalence classes:

$$U/P = \{\{x_1, x_4, x_5\}, \{x_2, x_8\}, \{x_3\}, \{x_6, x_7\}\}$$
$$U/Q = \{\{x_1, x_3, x_5\}, \{x_6\}, \{x_2, x_4, x_7, x_8\}\}$$
$$U/R = \{\{x_1, x_5\}, \{x_6\}, \{x_2, x_7, x_8\}, \{x_3, x_4\}\}$$

Thus, the relation $IND(\mathbf{R})$ has the equivalence classes:

$$U/IND(\mathbf{R}) = \{\{x_1, x_5\}, \{x_2, x_8\}, \{x_3\}, \{x_4\}, \{x_6\}, \{x_7\}\}.$$

The relation $P$ is indispensable in **R**, since

$$U/IND(\mathbf{R} - \{P\}) = \{\{x_1, x_3\}, \{x_2, x_7, x_8\}, \{x_3\}, \{x_4\}, \{x_6\}\} \neq U/IND(\mathbf{R})$$

For relation $Q$, we have:

$$U/IND(\mathbf{R} - \{Q\}) = \{\{x_1, x_3\}, \{x_2, x_8\}, \{x_3\}, \{x_4\}, \{x_6\}, \{x_7\}\} = U/IND(\mathbf{R})$$

thus the relation $Q$ is dispensable in **R**.

Similarly, for relation $R$, we have:

$$U/IND(\mathbf{R} - \{R\}) = \{\{x_1, x_3\}, \{x_2, x_8\}, \{x_3\}, \{x_4\}, \{x_6\}, \{x_7\}\} = U/IND(\mathbf{R})$$

hence the relation $R$ is also dispensable in **R**.

Thus, the classification defined by the set of three equivalence relations $P$, $Q$ and $R$ is the same as the classification defined by relation $P$ and $Q$ or $P$ and $R$.

To find reducts of the family $\mathbf{R} = \{P, Q, R\}$, we have to check whether pairs of relations $P, Q$ and $P, R$ are independent or not. Because $U/IND(\{P, Q\}) \neq U/IND(P)$ and $U/IND(\{P, Q\}) \neq U/IND(Q)$, the relations $P$ and $Q$ are independent. Consequently, $\{P, Q\}$ is a reduct of **R**. Similarly, we can see that $\{P, R\}$ is also a reduct of **R**.

Thus, there are two reducts of the family **R**, namely $\{P, Q\}$ and $\{P, R\}$, and $\{P, Q\} \cap \{P, R\} = \{P\}$ is the core of **R**.

The concepts of reduct and core defined above can be generalized. Let $P$ and $Q$ be equivalence relations over $U$. *P-positive region* of $Q$, denoted $POS_P(Q)$, is defined as follows:

$$POS_P(Q) = \bigcup_{X \in U/Q} \underline{P}X$$

The positive region of $Q$ is the set of all objects of the universe $U$ which can be properly classified to classes of $U/Q$ employing knowledge expressed by the classification $U/P$.

Let **P** and **Q** be families of equivalence relations over $U$. We say that $R \in \mathbf{P}$ is **Q**-dispensable in **P**, if

$$POS_{IND(\mathbf{P})}(IND(\mathbf{Q})) = POS_{IND(\mathbf{P}-\{R\})}(IND(\mathbf{Q}))$$

otherwise $R$ is **Q**-indispensable in **P**.

If every $R$ in **P** is **Q**-indispensable, we say that **P** is **Q**-independent. The family $\mathbf{S} \subseteq \mathbf{P}$ is called a **Q**-reduct of **P** iff **S** is the **Q**-independent subfamily of **P** and

$POS_{\mathbf{S}}(\mathbf{Q}) = POS_{\mathbf{P}}(\mathbf{Q})$. The set of all **Q**-indispensable elementary relation in **P** is called the **Q**-core, denoted $CORE_{\mathbf{Q}}(\mathbf{P})$.

The following proposition shows the relationship of relative reduct and core.

**Proposition 2.32** *$CORE_{\mathbf{Q}}(\mathbf{P}) = \bigcap RED_{\mathbf{Q}}(\mathbf{P})$, where $RED_{\mathbf{Q}}$ is the family of all **Q**-reducts of **P**.*

Let $POS_{\mathbf{P}}(\mathbf{Q})$ is the set of all objects which can be classified to elementary categories of knowledge **Q** employing knowledge **Q**. Knowledge **P** is **Q**-independent if the whole knowledge **P** is necessary to classify objects to elementary categories of knowledge **Q**.

The **Q**-core knowledge of knowldge **P** is the essential part of knowledge **P**, which cannot be eliminated without disturbing the ability to classify objects to elementary categories of **Q**.

The **Q**-reduct of knowledge **P** is the minimal subset of knowledge **P**, which provides the same classification of objects to elementary categories of knowledge **Q** as the whole knowledge **P**. Note that knowledge **P** can have more than one reduct.

Knowledge **P** with only one **Q**-reduct is, in a sense, deterministic, i.e., there is only one way of using elementary categories of knowledge **P** when classifying objects to elementary categories of knowledge **Q**.

If knowledge **P** has many **Q**-reducts, then it is non-deterministic, and there are in general many ways of using elementary categories of **P** when clasifying objects to elementary categories of **Q**.

This non-determinism is particularly strong if the core knowledge is void. But non-determinism introduces synonymy to the knowledge, which in some cases may be a drawback.

We turn to reduction of categories. Basic categories are pieces of knowledge, which can be considered as "building blocks" of concepts. Every concept in the knowledge base can be only expessed (exactly or approximately) in terms of basic categories.

On the other hand, every basic category is "build up" (is an intersection) of some elementary categories. Then, the question arises whether all the elementary categories are necessary to define the basic categories in question.

The problem can be formulated precisely as follows. Let $F = \{X_1, ..., X_n\}$ be a family of sets such that $X_i \subseteq U$. We say that $X_i$ is *dispensable*, if $\bigcap(F - \{X_i\}) = \bigcap F$; otherwise the set $X_i$ is *indispensable* in $F$.

The family $F$ is *independent* if all its complements are indispensable in $F$; otherwise $F$ is *dependent*. The family $H \subseteq F$ is a *reduct* of $F$, if $H$ is independent and $\bigcap H = \bigcap F$. The family of all indispensable sets in $F$ is called the *core* of $F$, denoted $CORE(F)$.

**Proposition 2.33** *$CORE(F) = \bigcap RED(F)$, where $RED(F)$ is the family of all reducts of $F$.*

Now, we introduce the example from Pawlak [1], Let the family of three sets be $F = \{X, Y, Z\}$, where

$X = \{x_1, x_3, x_8\}$

$Y = \{x_1, x_3, x_4, x_5, x_6\}$

$Z = \{x_1, x_3, x_4, x_6, x_7\}$.

Hence, $\bigcap F = X \cap Y \cap Z = \{x_1, x_3\}$. Bacause

$$\bigcap(F - \{X\}) = Y \cap Z = \{x_1, x_3, x_4, x_6\}$$

$$\bigcap(F - \{Y\}) = X \cap Z = \{x_1, x_3\}$$

$$\bigcap(F - \{Z\}) = X \cap Y = \{x_1, x_3\}$$

sets $Y$ and $Z$ are dispensable in the family $F$ and the family $F$ is dependent. Set $X$ iw the core of $F$. Families $\{X, Y\}$ and $\{X, Z\}$ are reducts of $F$ and $\{X, Y\} \cap \{X, Z\} = \{X\}$ is the core of $F$.

We also need a method to elminate superfluous categories from categories which are the union of some categories. The problem can be formulated in a way similar to the previous one, with the exception that now instead of intersection of sets we will need union of sets.

Let $F = \{X_1, ..., X_n\}$ be a family of sets such that $X_i \subseteq U$. We say that $X_i$ is *dispensable* in $\bigcup F$, if $\bigcup(F - |X_i\}) = \bigcup F$; otherwise the set $X_i$ is *indispensable* in $\bigcup F$.

The family $F$ is *independent* with respect to $\bigcup F$ if all its components are indispensable in $\bigcup F$; otherwise $F$ is *dependent* in $\bigcup F$. The family $H \subseteq F$ is a *reduct* of $\bigcup F$, if $H$ is independent with respect to $\bigcup H$ and $\bigcup H = \bigcup F$.

Here is th example from Pawlak [1]. Let $F = \{X, Y, Z, T\}$, where

$X = \{x_1, x_3, x_8\}$

$Y = \{x_1, x_2, x_4, x_5, x_6\}$

$Z = \{x_1, x_3, x_4, x_6, x_7\}$

$T = \{x_1, x_2, x_5, x_7\}$

Obviously, $\bigcup F = X \cup Y \cup Z \cup T = \{x_1, x_2, x_3, x_4, x_5, x_6, x_7, x_8\}$.

Because we have:

$$\bigcup(F - \{X\}) = \bigcup\{Y, Z, T\} = \{x_1, x_2, x_3, x_4, x_5, x_6, x_7\} \neq \bigcup F$$

$$\bigcup(F - \{Y\}) = \bigcup\{X, Z, T\} = \{x_1, x_2, x_3, x_4, x_5, x_6, x_7, x_8\} = \bigcup F$$

$$\bigcup(F - \{Z\}) = \bigcup\{X, Y, T\} = \{x_1, x_2, x_3, x_4, x_5, x_6, x_7, x_8\} = \bigcup F$$

$$\bigcup(F - \{T\}) = \bigcup\{X, Y, Z\} = \{x_1, x_2, x_3, x_4, x_5, x_6, x_7, x_8\} = \bigcup F$$

thus the only indispensable set in the family $F$ is the set $X$, and remaining sets $Y$, $Z$ and $T$ are dispensable in the family.

Hence, the following sets are reducts of $F$: $\{X, Y, Z\}$, $\{X, Y, T\}$, $\{X, Z, T\}$. That means that the concept $\bigcup F = X \cup Y \cup Z \cup T$, which is the union of $X$, $Y$, $Z$ and $T$ can be simplified and represented as union of smaller numbers of concepts.

Here, we discuss the relative reduct and the core of categories. Suppose that $F = \{X_1, ..., X_n\}$, $X_i \subseteq U$ and a subset $Y \subseteq U$ such that $\bigcap F \subseteq Y$.

We say that $X_i$ is *Y-dispensable* in $\bigcap F$, if $\bigcap(F - \{X_i\}) \subseteq Y$; otherwise the set $X_i$ is *Y-indispensable* in $\bigcap F$.

The family $F$ is *Y-independent* in $\bigcap F$, if all its complements are $Y$-indispensable in $\bigcap F$; otherwise $F$ is *Y-dependent* in $\bigcup F$.

The family $H \subseteq F$ is a *Y-reduct* of $\bigcap F$, if $H$ is $Y$-independenct in $\bigcap F$ otherwise $F$ and $\bigcap H \subseteq Y$.

The family of all $Y$-indispensable sets in $\bigcap F$ is called the *Y-core* of $F$, denoted $CORE_F(F)$. We also say that a $Y$-reduct ($Y$-core) is a relative reduct (core) with respect to $Y$.

**Proposition 2.34** *$CORE_Y(F) = \bigcap RED_Y(F)$, where $RED_Y(F)$ is the family of all $Y$-reducts of $F$.*

Thus, superfluous elementary categories can be eliminated from the basic categories in a similar way as the equivalence relations.

As discussed above, reduction of knowledge is to remove superfluous partitions (equivalence relations). For this task, the concept of reduct and core play significant roles.

## 2.7 Knowledge Representation

In this section, we discuss a *knowledge representation system* (KR system), which can be seen as a formal language. It is interpreted as data table and plays an important role in practical applications.

A knowledge representation system is a pair $S = (U, A)$, where $U$ is a non-empty finite set called *universe* and $A$ is a non-empty finite set of *primitive attributes*. Every primitive attribute $a \in A$ is a total function $a : U \rightarrow V_a$, where $V_a$ is the set of values of $a$, called the *domain* of $a$.

For every subset of attributes $B \subseteq A$, we associate a binary relation *IND*($B$), called an *indiscernibility relation*, defined as:

$$IND(B) = \{(x, y) \in U^2 \mid for\ every\ a \in B, a(x) = a(y)\}$$

Obviously, *IND*($B$) is an equivalence relation and the following holds.

$$IND(B) = \bigcap_{a \in B} IND(a)$$

Every subset $B \subseteq A$ is called an *attribute*. If $B$ is a single element set, then $B$ is called *primitive*, otherwise *compound*.

Attribute $B$ may be considered as a name of the relation $IND(B)$, or in other words, a name of knowledge represented by an equivalence relation $IND(B)$.

Thus, the knowledge representation system $S = (U, A)$ may be viewed as a description of a knowledge base $K = (U, \mathbf{R})$. Here, each equivalence relation in the knowledge base is represented by an attribute and each equivalence class of the relation by an attribute value.

It is noted that there is a one-to-one correspondence between knowledge bases and knowledge representation systems. To check it, it suffices to assign to arbitrary knowledge base $K = (U, \mathbf{R})$ a knowledge representation system $S = (U, A)$ in the following way.

If $R \in \mathbf{R}$ and $U/R = \{X_1, ..., X_k\}$, then to the set of attributes $A$ every attribute $a_R : U \rightarrow V_{a_R}$ such that $V_{a_R} = \{1, ..., k\}$ and $a_R(x) = i$ iff $x \in X$ for $i = 1, .., k$. Then, all notions of knowledge bases can be expressed in terms of notions of knowledge representation systems.

Consider the following knowledge representation system from Pawlak [1]:

Here, the universe $U$ consists of 8 elements numbered 1, 2, 3, 4, 5, 6, 7 and 8, the set of attributes is $A = \{a, b, c, d, e\}$, whereas $V = V_a = V_b = V_c = V_d = V_e = \{0, 1, 2\}$.

In Table 2.4, elements 1, 4 and 5 of $U$ are indiscernible by atribute $a$, elements 2, 7 and 8 are indiscernible by the set of attriutes $\{b, c\}$, and elements 2 and 7 are indiscernible by the set of attributes $\{d, e\}$.

Partitions generated by attributes in this system are given below:

$U/IND\{a\} = \{\{2, 8\}, \{1, 4, 5\}, \{3, 6, 7\}\}$
$U/IND\{b\} = \{\{1, 3, 5\}, \{2, 4, 7, 8\}, \{6\}\}$
$U/IND\{c, d\} = \{\{1\}, \{3, 6\}, \{2, 7\}, \{4\}, \{5\}, \{8\}\}$
$U/IND\{a, b, c\} = \{\{1, 5\}, \{2, 8\}, \{3\}, \{4\}, \{6\}, \{7\}\}$

**Table 2.4** KR-system 3

| $U$ | $a$ | $b$ | $c$ | $d$ | $e$ |
|---|---|---|---|---|---|
| 1 | 1 | 0 | 2 | 2 | 0 |
| 2 | 0 | 1 | 1 | 1 | 2 |
| 3 | 2 | 0 | 0 | 1 | 1 |
| 4 | 1 | 1 | 0 | 2 | 2 |
| 5 | 1 | 0 | 2 | 0 | 1 |
| 6 | 2 | 2 | 0 | 1 | 1 |
| 7 | 2 | 1 | 1 | 1 | 2 |
| 8 | 0 | 1 | 1 | 0 | 1 |

For example, for the set of attributes $C = \{a, b, c\}$ and the subset $X = \{1, 2, 3, 4, 5\}$ of the universe, we have $\underline{C}X = \{1, 2, 3, 4, 5\}, \overline{C}X = \{1, 2, 3, 4, 5, 8\}$ and $BN_C(X) = \{2, 8\}$.

Thus, the set $X$ is rough with respect to the attribute $C$, which is to say that we are unable to decide whether elements 2 and 8 are members of $X$ or not, employing the set of attributes $C$. For the rest of the universe classification of elements using the set $C$ of attributes is possible.

The set of attributes $C = \{a, b, c\}$ is dependent. The attributes $a$ and $b$ are indispensable, whereas the attribute $c$ is superfluous. Here, the dependency $\{a, b\} \Rightarrow \{c\}$ holds. Because $IND\{a, b\}$ has the blocks $\{1, 5\}, \{2, 8\}, \{3\}, \{4\}, \{6\}, \{7\}$, and $IND\{c\}$ has the blocks $\{1, 5\}, \{2, 7, 8\}, \{3, 4, 6\}, IND\{a, b\} \subset IND\{c\}$.

We next compute the degree of dependency of attribute $D = \{d, e\}$ from the attributes $C = \{a, b, c\}$ in Table 2.4. The partition $U/IND(C)$ consists of the blocks, $X_1 = \{1\}, X_2 = \{2, 7\}, X_3 = \{3, 6\}, X_4 = \{4\}, X_5 = \{5, 8\}$. The partition $U/IND(D)$ consists of the blocks, $Y_1 = \{1, 5\}, Y_2 = \{2, 8\}, Y_3 = \{3\}, Y_4 = \{4\}, Y_5 = \{6\}, Y_6 = \{7\}$.

Because $\underline{C}X_1 = \emptyset, \underline{C}X_2 = Y_6, \underline{C}X_3 = Y_3 \cup Y_5, \underline{C}X_4 = Y_4$ and $\underline{C}X_5 = \emptyset$, we have $POS(D) = Y_3 \cup Y_4 \cup Y_5 \cup Y_6 = \{3, 4, 6, 7\}$.

Namely, only these elements can be classified into blocks of the partition $U/IND(D)$ employing the set $C = \{a, b, c\}$ attributes. Hence, the degree of dependency between $C$ and $D$ is $\gamma_C(D) = 4/8 = 0.5$.

The set of attributes $C$ is $D$-independent, and the attribute $a$ is $D$-indispensable. This means that the the $D$-core of $C$ is one attribute set $\{a\}$. Thus, there are the following dependencies: $\{a, b\} \Rightarrow \{d, e\}$ and $\{a, c\} \Rightarrow \{d, e\}$ in the table.

When speaking about attributes, it is obvious that they may have varying importance in the analysis of considered issues. To find out the significance of a specific attribute (or group of attributes) it seems reasonable to drop the attribute from the table and see how classification will be changed without this attribute.

If removing the attribute will change the classification considerably it means that its significance is high-in the opposite case, the significance should be low. The idea can be precisely employing the concept of a positive region.

As a measure of the signignificance of the subset of attributes $B' \subseteq B$ with respect to the classification induced by a set of attributes $C$, we will mean the difference:

$$\gamma_B(C) - \gamma_{B-B'}(C)$$

which expresses how the positive region of the classification $U/IND(C)$ when classifying the object by means of attribute $B$ will be affect if we drop some attributes (subset $B'$) from the set $B$.

Let us compute the significance of the attributes $a, b$ and $c$ with respect to the set of attributes $\{d, e\}$ in Table 2.4. $POS_C(D) = \{3, 4, 6, 7\}$, where $C = \{a, b, c\}$ and $D = \{d, e\}$. Because

$$U/IND\{b, c\} = \{\{1, 5\}, \{2, 7, 8\}, \{3\}, \{4\}, \{6\}\}$$
$$U/IND\{a, c\} = \{\{1, 5\}, \{2, 8\}, \{3, 6\}, \{4\}, \{7\}\}$$
$$U/IND\{a, b\} = \{\{1, 5\}, \{2, 8\}, \{3\}, \{4\}, \{6\}, \{7\}\}$$
$$U/IND\{d, e\} = \{\{1\}, \{2, 7\}, \{3, 6\}, \{4\}, \{5, 8\}\}$$

we have:

$$POS_{C-\{a\}}(D) = \{3, 4, 6\}$$
$$POS_{C-\{b\}}(D) = \{3, 4, 6, 7\}$$
$$POS_{C-\{c\}}(D) = \{3, 4, 6, 7\}$$

Consequently, corresponding accuracies are:

$$\gamma_{C-\{a\}}(D) = 0.125,$$
$$\gamma_{C-\{b\}}(D) = 0,$$
$$\gamma_{C-\{c\}}(D) = 0.$$

Thus, the attribute $a$ is most significant, since it most changes the positive region of $U/IND(D)$, i.e., without the attribute $a$ we are unable to classify object 7 to classes of $U/IND(D)$.

Note that the attribute $a$ is $D$-indispensable and the attributes $b$ and $c$ are $D$-dispensable. Thus, the attribute $a$ is the core of $C$ with respect to $D$ ($D$-core of $C$) and $\{a, b\}$ and $\{a, c\}$ are reducts of $C$ with respect to $D$ ($D$-reducts of $C$).

Knowledge representation systems can be expressed by means of tables, but as will be discussed in Chap. 4 it can be also formalized in the framework of modal logic.

We may find some similarities between knowledge representation systems and *relational databases* (cf. Codd [15]) in that the concept of table plays a crucial role. There is, however, an essential difference between these two models.

Most importantly, the relational model is not interested in the meaning of the information stored in the table. It focusses on efficient data structuring and manipulation. Consequently, the objects about which information is contained in the table are not represented in the table.

On the other hand, in knowledge representation system, all objects are explicitly represented and the attribute values, i.e., the table entries, have associated explicit meaning as features or properties of the objects.

## 2.8 Decision Tables

*Decision tables* can be seen as a special, important class of knowledge representation systems, and can be used for applications. Let $K = (U, A)$ be a knowledge representation system and $C, D \subset A$ be two subsets of attributes, called *condition* and *decision* attributes, respectively.

KR-system with distinguished condition and decision attributes is called a *decision table*, denoted $T = (U, A, C, D)$ or in short DC. Equivalence clases of the relations $IND(C)$ and $IND(D)$ are called *condition* and *decision classes*, respectively.

With every $x \in U$, we associate a function $d_x : A \rightarrow V$, such that $d_x(a) = a(x)$ for every $a \in C \cup D$; the function $d_x$ is called a *decision rule* (in $T$), and $x$ is referred as a *label* of the decision rule $d_x$.

Note that elements of the set $U$ in a decision table do not represent in general any real objects, but are simple identifiers of decision rules.

If $d_x$ is a decision rule, then the restriction of $d_x$ to $C$, denoted $d_x \mid C$, and the restriction of $d_x$ to $D$, denoted $d_x \mid D$ are called *conditions* and *decisions* (actions) of $d_x$, respectively.

The decision rule $d_x$ is *consistent* (in $T$) if for every $y \neq x$, $d_x \mid C = d_y \mid C$ implies $d_x \mid D = d_y \mid D$; otherwise the decision rule is *inconsistent*.

A decision table is *consistent* if all its decision rules are consistent; otherwise the decision table is *inconsistent*. Consistency (inconsistency) sometimes may be interpreted as determinism (non-determinism).

**Proposition 2.35** *A decision table $T = (U, A, C, D)$ is consistent iff $C \Rightarrow D$.*

From Proposition 2.35, it follows that the practical method of checking consistency of a decision table is by simply computing the degree of dependency between condition and decision attributes. If the degree of dependency equals to 1, then we conclude that the table is consistent; otherwise it is inconsistent.

**Proposition 2.36** *Each decision table $T = (U, A, C, D)$ can be uniquely decomposed into two decision tables $T_1 = (U, A, C, D)$ and $T_2 = (U, A, C, D)$ such that $C \Rightarrow_1 D$ in $T_1$ and $C \Rightarrow_0 D$ in $T_2$ such that $U_1 = POS_C(D)$ and $U_2 = \bigcup_{X \in U/IND(D)} BN_C(X)$.*

Proposition 2.36 states that we can decompose the table into two subtables; one totally inconsistent with dependency coefficient equal to 0, and the second entirely consistent with the dependency equal to 1. This decomposition however is possible only if the degree of dependency is greater than 0 and different from 1.

Consider Table 2.5 from Pawlak [1].

Assume that $a$, $b$ and $c$ are condition attributes, and $d$ and $e$ are decision attributes. In this table, for instance, the decision rule 1 is inconsistent, whereas the decision rule 3 is consistent. By Proposition 2.36, we can decompose Decision Table 1 into the following two tables:

Table 2.2 is consistent, whereas Table 2.3 is totally inconsistent, which mens all decision rules in Table 2.2 are consistent, and in Table 2.3 all decision rules are inconsistent.

Simplification of decision tables is very important in many applications, e.g. software engineering. An example of simplification is the redeuction of condition attributes in a decision table.

**Table 2.5** Decision Table 1

| $U$ | $a$ | $b$ | $c$ | $d$ | $e$ |
|---|---|---|---|---|---|
| 1 | 1 | 0 | 2 | 2 | 0 |
| 2 | 0 | 1 | 1 | 1 | 2 |
| 3 | 2 | 0 | 0 | 1 | 1 |
| 4 | 1 | 1 | 0 | 2 | 2 |
| 5 | 1 | 0 | 2 | 0 | 1 |
| 6 | 2 | 2 | 0 | 1 | 1 |
| 7 | 2 | 1 | 1 | 1 | 2 |
| 8 | 0 | 1 | 1 | 0 | 1 |

In the reduced decision table, the same decisions can be based on a smaller number of conditions. This kind of simplification eliminates the need for checking unnecessary conditions.

Pawlak proposed simplification of decision tables which includes the following steps:

(1) Computation of reducts of condition attributes which is equivalent to elimination of some column from the decision table.
(2) Elimination of duplicate rows.
(3) Elimination of superfluous values of attributes.

Thus, the method above consists in removing superfluous condition attributes (columns), duplicate rows and, in addition to that, irrelevant values of condition attributes.

By the above procedure, we obtain an "incomplete" decision table, containing only those values of condition attributes which are necessary to make decisions. According to our definition of a decision table, the incomplete table is not a decision table and can be treated as an abbreviation of such a table.

For the sake of simplicity, we assume that the set of condition attribute is already reduced, i.e., there are not superfluous condition attributes in the decision table.

With every subset of attributes $B \subseteq A$, we can associate partition $U/IND(B)$ and consequently the set of condition and decision attributes define partitions of objects into condition and decision classes.

We know that with every subset of attributes $B \subseteq A$ and object $x$ we may associate set $[x]_B$, which denotes an equivalence class of the relation $IND(B)$ containing an object $x$, i.e., $[x]_B$ is an abbreviation of $[x]_{IND(B)}$.

Thus, with any set of condition attributes $C$ in a decision rule $d_x$ we can associate set $[x]_C = \bigcap_{a \in C}[x]_a$. But, each set $[x]_a$ is uniquely determined by attribute value $a(x)$. Hence, in order to remove superfluous values of condition attributes, we have to eliminate all superfluous equivalence classes $[x]_a$ from the equivalence class $[x]_C$. Thus, problems of elimination of superfluous values of attributes and elimination of corresponding equivalence classes are equivalent.

Consider the following decision table from Pawlak [1].

Here, $a$, $b$ and $c$ are condition attributes and $e$ is a decision attribute.

It is easy to compute that the only $e$-dispensable condition attribute is $c$; consequently, we can remove column $c$ in Table 2.4, which yields Table 2.5:

In the next step, we have to reduce superfluous values of condition attributes in every decision rule. First, we have to compute core values of condition attributes in every decision rule.

Here, we compute the core values of condition attributes for the first decision rule, i.e., the core of the family of sets

$$\mathbf{F} = \{[1]_a, [1]_b, [1]_d\} = \{\{1, 2, 4, 5\}, \{1, 2, 3\}, \{1, 4\}\}$$

From this we have:

$$[1]_{\{a,b,d\}} = [1]_a \cap [1]_b \cap [1]_d = \{1, 2, 4, 5\} \cap \{1, 2, 3\} \cap \{1, 4\} = \{1\}.$$

Moreover, $a(1) = 1, b(1) = 0$ and $d(1) = 1$. In order to find dispensable categories, we have to drop one category at a time and check whether the intersection of remaining categories is still included in the decision category $[1]_e = \{1, 2\}$, i.e.,

$$[1]_b \cap [1]_d = \{1, 2, 3\} \cap \{1, 4\} = \{1\}$$
$$[1]_a \cap [1]_d = \{1, 2, 4, 5\} \cap \{1, 4\} = \{1, 4\}$$
$$[1]_a \cap [1]_b = \{1, 2, 4, 5\} \cap \{1, 2, 3\} = \{1, 2\}$$

This means that the core value is $b(1) = 0$. Similarly, we can compute remaining core values of condition attributes in every decision rule and the final results are represented in Table 2.6.

Then, we can proceed to compute value reducts. As an example, let us compute value reducts for the first decision rule of the decision table.

Accordingly to the definition of it, in order to compute reducts of the family $\mathbf{F} = \{[1]_a, [1]_b, [1]_d\} = \{\{1, 2, 3, 5\}, \{1, 2, 3\}, \{1, 4\}\}$, we have to find all subfamilies $\mathbf{G} \subseteq \mathbf{F}$ such that $\bigcap \mathbf{G} \subseteq [1]_e = \{1, 2\}$. There are four following subfamilies of $\mathbf{F}$:

$$[1]_b \cap [1]_d = \{1, 2, 3\} \cap \{1, 4\} = \{1\}$$
$$[1]_a \cap [1]_d = \{1, 2, 4, 5\} \cap \{1, 4\} = \{1, 4\}$$
$$[1]_a \cap [1]_b = \{1, 2, 4, 5\} \cap \{1, 2, 3\} = \{1\}$$

**Table 2.6** Decision Table 2.2

| $U$ | $a$ | $b$ | $c$ | $d$ | $e$ |
|---|---|---|---|---|---|
| 3 | 2 | 0 | 0 | 1 | 1 |
| 4 | 1 | 1 | 0 | 2 | 2 |
| 6 | 2 | 2 | 0 | 1 | 1 |
| 7 | 2 | 1 | 1 | 1 | 2 |

and only two of them

$$[1]_b \cap [1]_d = \{1, 2, 3\} \cap \{1, 4\} = \{1\} \subseteq [1]_e = \{1, 2\}$$
$$[1]_a \cap [1]_b = \{1, 2, 4, 5\} \cap \{1, 2, 3\} = \{1\} \subseteq [1]_e = \{1, 2\}$$

are reducts of the family **F**. Hence, we have two values reducts: $b(1) = 0$ and $d(1) = 1$ or $a(1) = 1$ and $b(1) = 0$. This means that the attribute values of attributes $a$ and $b$ or $d$ and $e$ are characteristic for decision class 1 and do not occur in any other decision classes in the decision table. We see also that the value of attribute $b$ is the intersection of both value reducts, $b(1) = 0$, i.e., it is the core value.

In Table 2.7, we list value reducts fir all decision rules in Table 2.1.

Seen from Decision Table 2.7, for decision rules 1 and 2 we have two value reducts of condition attributes. Decision rules 3,4 and 5 have only one value reducts of condition attributes for each decision rule row. The remaining decision rules 6 and 7 contain two and three value reducts, respectively.

Hence, there are two reduced form of decision rule 1 and 2, decision rule 3, 4 and 5 have only one reduced form each, decision rule 6 has two reducts and decision rule 7 has three reducts.

Thus, there are $4 \times 2 \times 3 = 24$ (not necessarily different) solutions to our problem. One such solution is presented in Decision Table 2.8.

Another solution is shown in Decision Table 2.9.

Because decision rules 1 and 2 are identical, and so are rules 5, 6 and 7, we can represent Decision Table 2.10:

In fact, enumeration of decision rules is not essential, so we can enumerate them arbitrary and we get as a final result Decision Table 2.11:

**Table 2.7** Decision Table 2.3

| $U$ | $a$ | $b$ | $c$ | $d$ | $e$ |
|---|---|---|---|---|---|
| 1 | 1 | 0 | 2 | 2 | 0 |
| 2 | 0 | 1 | 1 | 1 | 2 |
| 5 | 1 | 0 | 2 | 0 | 1 |
| 8 | 0 | 1 | 1 | 0 | 1 |

**Table 2.8** Decision Table 2.4

| $U$ | $a$ | $b$ | $c$ | $d$ | $e$ |
|---|---|---|---|---|---|
| 1 | 1 | 0 | 0 | 1 | 1 |
| 2 | 1 | 0 | 0 | 0 | 1 |
| 3 | 0 | 0 | 0 | 0 | 0 |
| 4 | 1 | 1 | 0 | 1 | 0 |
| 5 | 1 | 1 | 0 | 2 | 2 |
| 6 | 2 | 1 | 0 | 2 | 2 |
| 7 | 2 | 2 | 2 | 2 | 2 |

**Table 2.9** Decision Table 2.5

| $U$ | $a$ | $b$ | $d$ | $e$ |
|---|---|---|---|---|
| 1 | 1 | 0 | 1 | 1 |
| 2 | 1 | 0 | 0 | 1 |
| 3 | 0 | 0 | 0 | 0 |
| 4 | 1 | 1 | 1 | 0 |
| 5 | 1 | 1 | 2 | 2 |
| 6 | 2 | 1 | 2 | 2 |
| 7 | 2 | 2 | 2 | 2 |

**Table 2.10** Decision Table 2.6

| $U$ | $a$ | $b$ | $d$ | $e$ |
|---|---|---|---|---|
| 1 | – | 0 | – | 1 |
| 2 | 1 | – | – | 1 |
| 3 | 0 | – | – | 0 |
| 4 | – | 1 | 1 | 0 |
| 5 | – | – | 2 | 2 |
| 6 | – | – | – | 2 |
| 7 | – | – | – | 2 |

**Table 2.11** Decision Table 2.7

| $U$ | $a$ | $b$ | $d$ | $e$ |
|---|---|---|---|---|
| 1 | 1 | 0 | × | 1 |
| $1'$ | × | 0 | 1 | 1 |
| 2 | 1 | 0 | × | 1 |
| $2'$ | 1 | × | 0 | 1 |
| 3 | 0 | × | × | 0 |
| 4 | × | 1 | 1 | 0 |
| 5 | × | × | 2 | 2 |
| 6 | × | × | 2 | 2 |
| $6'$ | 2 | × | × | 2 |
| 7 | × | × | 2 | 2 |
| $7'$ | × | 2 | × | 2 |
| $7''$ | 2 | × | × | 2 |

This solution is referred to as *minimal*. The presented method of decision table simplification can be named *semantic*, since it refers to the meaning of the information contained in the table. Another decision table simplification is also possible and can be named *syntactic*. It is described within the framework of decision logic (Table 2.12).

To simplify a decision table, we should first find reducts of condition attributes, remove duplicate rows and then find value-reducts of condition attributes and again, if necessary, remove duplicate rows (Table 2.13).

This method leads to a simple algorithm for decision table simplification, Note that a subset of attributes may have more than one reduct (relative reduct). Thus, the simplification of decision table does not yield unique results. Some decision tables possibly can be optimized according to preassumed criteria (Table 2.14).

**Table 2.12** Decision Table 2.8

| $U$ | $a$ | $b$ | $d$ | $e$ |
|---|---|---|---|---|
| 1 | 1 | 0 | × | 1 |
| 2 | 1 | × | 0 | 1 |
| 3 | 0 | × | × | 0 |
| 4 | × | 1 | 1 | 0 |
| 5 | × | × | 2 | 2 |
| 6 | × | × | 2 | 2 |
| 7 | 2 | × | × | 2 |

**Table 2.13** Decision Table 2.9

| $U$ | $a$ | $b$ | $d$ | $e$ |
|---|---|---|---|---|
| 1 | 1 | 0 | × | 1 |
| 2 | 1 | 0 | × | 1 |
| 3 | 0 | × | × | 0 |
| 4 | × | 1 | 1 | 0 |
| 5 | × | × | 2 | 2 |
| 6 | × | × | 2 | 2 |
| 7 | × | × | 2 | 2 |

**Table 2.14** Decision Table 2.10

| $U$ | $a$ | $b$ | $d$ | $e$ |
|---|---|---|---|---|
| 1, 2 | 1 | 0 | × | 1 |
| 3 | 0 | × | × | 0 |
| 4 | × | 1 | 1 | 0 |
| 5, 6, 7 | × | × | 2 | 2 |

**Table 2.15** Decision Table 2.11

| $U$ | $a$ | $b$ | $d$ | $e$ |
|---|---|---|---|---|
| 1 | 1 | 0 | × | 1 |
| 2 | 0 | × | × | 0 |
| 3 | × | 1 | 1 | 0 |
| 4 | × | × | 2 | 2 |

We have finished the presentation of some topics in rough set theory. Pawlak also established other formal results about rough sets and discusses advantages of rough set theory. We here omit these issues; see Pawlak [1] (Table 2.15).

## References


1. Pawlak, P.: Rough Sets: Theoretical Aspects of Reasoning about Data. Kluwer, Dordrecht (1991)
2. Ziarko, W.: Variable precision rough set model. J. Comput. Syst. Sci. **46**, 39–59 (1993)
3. Pawlak, P.: Rough sets. Int. J. Comput. Inf. Sci. **11**, 341–356 (1982)
4. Shen, Y., Wang, F.: Variable precision rough set model over two universes and its properties. Soft. Comput. **15**, 557–567 (2011)
5. Zadeh, L.: Fuzzy sets. Inf. Control **8**, 338–353 (1965)
6. Zadeh, L.: Fuzzy sets as a basis for a theory of possibility. Fuzzy Sets Syst. **1**, 3–28 (1976)
7. Negoita, C., Ralescu, D.: Applications of Fuzzy Sets to Systems Analysis. Wiley, New York (1975)
8. Nakamura, A., Gao, J.: A logic for fuzzy data analysis. Fuzzy Sets Syst. **39**, 127–132 (1991)
9. Quafafou, M.: $\alpha$-RST: a generalizations of rough set theory. Inf. Sci. **124**, 301–316
10. Cornelis, C., De Cock, J., Kerre, E.: Intuitionistic fuzzy rough sets: at the crossroads of imperfect knowledge. Expert Syst. **20**, 260–270 (2003)
11. Atnassov, K.: Intuitionistic Fuzzy Sets. Physica, Haidelberg (1999)
12. Ganter, B., Wille, R.: Formal Concept Analysis. Springer, Berlin (1999)
13. Ore, O.: Galois connexion. Trans. Am. Math. Soc. **33**, 493–513 (1944)
14. Armstrong, W.: Dependency structures in data base relationships, IFIP'74, pp. 580–583 (1974)
15. Codd, E.: A relational model of data for large shared data banks. Commun. ACM **13**, 377–387 (1970)

# Chapter 3
# Non-classical Logics

**Abstract** This chapter surveys some non-classical logics. They are closely related to the foundations of rough set theory. We provide the basics of modal, many-valued, intuitionistic and paraconsistent logic.

## 3.1 Modal Logic

*Non-classical logic* is a logic which differs from classical logic in some points. There are many systems of non-classical logic in the literature. Some non-classical logics are closely tied with foundations of rough set theory.

There are two types of non-classical logics. The first type is considered as an *extension* of classical logic. It extends classical logic with new features. For instance, modal logic adds modal operators to classical logic.

The second type is an *alternative* (or rival) to classical logic. It therefore denies some of the features of classical logic. For example, many-valued logic is based on many truth-values, whereas classical logic uses two truth-values, i.e. true and false.

These two types of non-classical logics are conceptually different and their uses heavily depend on applications. In some cases, they can provide more promising results than classical logic. In the following, we provide the basics of modal, many-valued, intuitionistic, and paraconsistent logic.

*Modal logic* extends classical logic with modal operators to represent *intensional concepts*. Intensional concepts are beyond the scope of truth and falsity. So new mechanism for intensionality should be devised.

The role can be played by a modal operator. Generally, $\Box$ (necessity) and $\Diamond$ (possibility) are used as modal operators. A formula of the form $\Box A$ reads "$A$ is necessarily true" and $\Box$ "$A$ is possibly true", respectively. These are dual in the sense that $\Box A \leftrightarrow \neg\Diamond\neg A$.

Reading modal operators differently, we can obtain other intensional logics capable of formalizing some intensional concepts. Currently, many variants of modal logics are known. We list some of them as follows: tense logic, epistemic logic, doxastic logic, deontic logic, dynamic logic, intensional logic, etc.



S. Akama et al., *Reasoning with Rough Sets*, Intelligent Systems Reference Library 142, https://doi.org/10.1007/978-3-319-72691-5_3

Now, we present proof and model theory for modal logic. The language of the minimal modal logic denoted **K** is the classical propositional logic **CPC** with the necessity operator $\Box$. The name "K" is after Kripke.

A Hilbert system for **K** is formalized as follows:

**Modal Logic K**
**Axiom**
(CPC) Axioms of **CPC**
(K) $\Box(A \rightarrow B) \rightarrow (\Box A \rightarrow \Box B)$
**Rules of Inference**
(MP) $\vdash A, \vdash A \rightarrow B \Rightarrow \vdash B$
(NEC) $\vdash A \Rightarrow \vdash \Box A$

Here, $\vdash A$ means that $A$ is provable in **K**. (NEC) is called the *necessitation*. The notion of proof is defined as usual.

Systems of *normal modal logic* can be obtained by adding extra axioms which describe properties of modality. Some of the important axioms are listed as follows:

(D) $\Box A \rightarrow \Diamond A$
(T) $\Box A \rightarrow A$
(B) $A \rightarrow \Box\Diamond A$
(4) $\Box A \rightarrow \Box\Box A$
(5) $\Diamond A \rightarrow \Box\Diamond A$

The name of normal modal logic is systematically given by the combination of axioms. For instance, the extension of **K** with the axiom (D) is called KD. However, some such systems traditionally have the following names:

**D** = KD
**T** = KT
**B** = KB
**S4** = KT4
**S5** = KT5

Before the 1960s, the study of modal logic was mainly proof-theoretical due to the lack of model theory. A semantics of modal logic has been developed by Kripke and it is now called *Kripke semantics*; see Kripke [1–3].

Kripke semantics uses a *possible world* to interpret modal operators. Intuitively, the interpretation of $\Box A$ says that $A$ is true in all possible worlds. Possible worlds are linked with the actual world by means of the *accessibility relation*.

A *Kripke model* for the normal modal logic **K** is defined as a triple $M = \langle W, R, V \rangle$, where $W$ is a non-empty set of possible worlds, $R$ is an accessibility relation on $W \times W$, and $V$ is a valuation function: $W \times PV \rightarrow \{0, 1\}$. We here denote by $PV$ a set of propositional variables. $F = \langle W, R \rangle$ is called a *frame*.

We write $M, w \models A$ to mean that a formula $A$ is true at a world $w$ in the model $M$. Let $p$ be a propositional variable and *false* be absurdity. Then, $\models$ can be defined as follows:

$M, w \models p \Leftrightarrow V(w, p) = 1$
$M, w \not\models false$
$M, w \models \neg A \Leftrightarrow M, w \not\models A$
$M, w \models A \wedge B \Leftrightarrow M, w \models A$ and $M, w \models B$
$M, w \models A \vee B \Leftrightarrow M, w \models A$ or $M, w \models B$
$M, w \models A \rightarrow B \Leftrightarrow M, w \models A \Rightarrow M, w \models B$
$M, w \models \Box A \Leftrightarrow \forall v(wRv \Rightarrow M, v \models A)$
$M, w \models \Diamond A \Leftrightarrow \exists v(wRv$ and $M, v \models A)$

Here, there are no restrictions on the property of $R$. We say that a formula $A$ is *valid* in the modal logic $S$, written $M \models_S A$, just in case $M, w \models A$ for every world $w$ and every model $M$.

We know that the minimal modal logic **K** is complete.

**Theorem 3.1** $\vdash_{\mathbf{K}} A \Leftrightarrow \models_{\mathbf{K}} A$.

By imposing some restrictions on the accessibility relation $R$, we can give Kripke models for various normal modal logics. The correspondences of axioms and conditions on $R$ are given as follows:

| Axiom | Conditions on $R$ |
|---|---|
| (K) | No conditions |
| (D) | $\forall w \exists v(wRv)$ (serial) |
| (T) | $\forall w(wRw)$ (reflexive) |
| (4) | $\forall wvu(wRv$ and $vRu \Rightarrow wRu)$ (transitive) |
| (5) | $\forall wvu(wRv$ and $wRu \Rightarrow vRu)$ (euclidean) |

For example, the accessibility relation in a Kripke model for modal logic **S4** is reflexive ad transitive since it needs axioms (K), (T) and (4). The completeness results of several modal logics have been established; see Hughes and Cresswell [4] for details.

If we read modal operators differently, then other types of modal logics listed above can be obtained. These logics can deal with various problems, and modal logic is of special importance to applications.

## 3.2 Many-Valued Logic

*Many-valued logic*, also known as multiple-valued logic, is a family of logics which have more than two truth-values. Namely, many-valued logics can express other possibilities in addition to truth and falsity.

The idea of many-valued logic is implicit in Aristotle's thinking concerning *future contingents*. Now, many-valued logics are widely used to treat problems in various areas. In particular, three-valued and four-valued logics are well known for applications in computer science. It is also noted that the so-called *fuzzy logic* is classified as a many-valued (infinite-valued) logic.

We start with the exposition of *three-valued logic*. The first serious attempt to formalize a three-valued logic has been done by Łukasiewicz in [5]. His system is now known as Łukasiewicz's three-valued logic, denoted $\mathbf{L_3}$, in which the third truth-valued reads "indeterminate" or "possible".

Łukasiewicz considered that future contingent propositions should receive the third truth-value denoted by $I$, which is neither true nor false, although his interpretation is controversial.

The language of $\mathbf{L_3}$ comprises conjunction ($\wedge$), disjunction ($\vee$), implication ($\rightarrow_L$) and negation ($\sim$). We will omit the subscript $L$ when the context is clear. The semantics for many-valued logics can be usually given by using the truth-value tables. The truth-value tables for $\mathbf{L_3}$ are as follows:

Here, we should note that both the law of excluded middle $A \vee \sim A$ and the law of non-contradiction $\sim (A \wedge \sim A)$, which are important principles of classical logic, do not hold. In fact, these receive $I$ when the truth-values of compound formulas are $I$ (Table 3.1).

A Hilbert system for $\mathbf{L_3}$ is as follows:

**Lukasiewicz's Three-Valued Logic $\mathbf{L_3}$**
**Axiom**
(L1) $A \rightarrow (B \rightarrow A)$
(L2) $(A \rightarrow B) \rightarrow ((B \rightarrow C) \rightarrow (A \rightarrow C))$
(L3) $((A \rightarrow \sim A) \rightarrow A) \rightarrow A$

**Table 3.1** Truth-value tables of $\mathbf{L_3}$

| $A$ | $\sim A$ |
|---|---|
| $T$ | $F$ |
| $I$ | $I$ |
| $F$ | $T$ |

| $A$ | $B$ | $A \wedge B$ | $A \vee B$ | $A \rightarrow_L B$ |
|---|---|---|---|---|
| $T$ | $T$ | $T$ | $T$ | $T$ |
| $T$ | $F$ | $F$ | $T$ | $F$ |
| $T$ | $I$ | $I$ | $T$ | $I$ |
| $F$ | $T$ | $F$ | $F$ | $T$ |
| $F$ | $F$ | $F$ | $F$ | $T$ |
| $F$ | $I$ | $F$ | $I$ | $T$ |
| $I$ | $T$ | $I$ | $T$ | $T$ |
| $I$ | $F$ | $F$ | $I$ | $I$ |
| $I$ | $I$ | $I$ | $I$ | $T$ |

**Table 3.2** Truth-value tables of $\mathbf{K_3}$

| $A$ | $\sim A$ |
|---|---|
| $T$ | $F$ |
| $I$ | $I$ |
| $F$ | $T$ |

| $A$ | $B$ | $A \wedge B$ | $A \vee B$ | $A \rightarrow_K B$ |
|---|---|---|---|---|
| $T$ | $T$ | $T$ | $T$ | $T$ |
| $T$ | $F$ | $F$ | $T$ | $F$ |
| $T$ | $I$ | $I$ | $T$ | $I$ |
| $F$ | $T$ | $F$ | $F$ | $T$ |
| $F$ | $F$ | $F$ | $F$ | $T$ |
| $F$ | $I$ | $F$ | $I$ | $T$ |
| $I$ | $T$ | $I$ | $T$ | $T$ |
| $I$ | $F$ | $F$ | $I$ | $I$ |
| $I$ | $I$ | $I$ | $I$ | $I$ |

(L4) $(\sim A \rightarrow \sim B) \rightarrow (B \rightarrow A)$
**Rules of Inference**
(MP) $\vdash A, \vdash A \rightarrow B \Rightarrow \vdash B$

Here, $\wedge$ and $\vee$ are defined by means of $\sim$ and $\rightarrow_L$ in the following way.

$A \vee B =_{\text{def}} (A \rightarrow B) \rightarrow B$
$A \wedge B =_{\text{def}} \sim (\sim A \vee \sim B)$

Kleene also proposed three-valued logic $\mathbf{K_3}$ in connection with recursive function theory; see Kleene [6]. $\mathbf{K_3}$ differs from $\mathbf{L_3}$ in its interpretation of implication $\rightarrow_K$. The truth-value tables of $\mathbf{K_3}$ are given as Table 3.2.

In $\mathbf{K_3}$, the third truth-value reads "undefined". Consequently, $\mathbf{K_3}$ can be applied to theory of programs. There are no tautologies in $\mathbf{K_3}$, thus implying that we cannot provide a Hilbert system for it.

$\mathbf{K_3}$ is usually called *Kleene's strong three-valued logic*. In the literature, *Kleene's weak three-valued logic* also appears, in which a formula evaluates as $I$ if any compound formula evaluates as $I$. Kleene's weak three-valued logic is equivalent to Bochvar's three-valued logic.

*Four-valued logic* is suited as a logic for a computer which must deal with incomplete and inconsistent information. Belnap introduced a four-valued logic which can formalize the internal states of a computer; see Belnap [7, 8].

There are four states, i.e. ($T$), ($F$), (*None*) and (*Both*), to recognize an input in a computer. Based on these states, a computer can compute suitable outputs.

---

(*T*) a proposition is true.
(*F*) a proposition is false.
(*N*) a proposition is neither true nor false.
(*B*) a proposition is both true and false.

---

Here, (*N*) and (*B*) abbreviate (*None*) and (*Both*), respectively. From the above, (*N*) corresponds to incompleteness and (*B*) inconsistency. Four-valued logic can be thus seen as a natural extension of three-valued logic.

In fact, Belnap's four-valued logic can model both incomplete information (*N*) and inconsistent information (*B*). Belnap proposed two four-valued logics **A4** and **L4**.

The former can cope only with atomic formulas, whereas the latter can handle compound formulas. **A4** is based on the *approximation lattice* depicted as Fig. 3.1.

Here, *B* is the least upper bound and *N* is the greatest lower bound with respect to the ordering $\leq$.

**L4** is based on the *logical lattice* depicted as Fig. 3.2.

**L4** has logical symbols; $\sim, \wedge, \vee$, and is based on a set of truth-values $\mathbf{4} = \{T, F, N, B\}$. One of the features of **L4** is the monotonicity of logical symbols. Let $f$ be a logical operation.

It is said that $f$ is monotonic iff $a \subseteq b \Rightarrow f(a) \subseteq f(b)$. To guarantee the monotonicity of conjunction and disjunction, they must satisfy the following:

$$a \wedge b = a \Leftrightarrow a \vee b = b$$
$$a \wedge b = b \Leftrightarrow a \vee b = a$$

The truth-vaue tables for **L4** are as follows (Table 3.3).

Belnap gave a semantics for the language with the above logical symbols. A *setup* is a mapping a set of atomic formulas *Atom* to the set **4**. Then, the meaning of formulas of **L4** are defined as follows:

**Fig. 3.1** Approximation lattice

```
    B
  /   \
T  A4  F
  \   /
    N
```

**Fig. 3.2** Logical lattice

```
    T
  /   \
N  L4  B
  \   /
    F
```

**Table 3.3** Truth-value tables of **L4**

| | $N$ | $F$ | $T$ | $B$ |
|---|---|---|---|---|
| $\sim$ | $B$ | $T$ | $F$ | $N$ |

| $\wedge$ | $N$ | $F$ | $T$ | $B$ |
|---|---|---|---|---|
| $N$ | $N$ | $F$ | $N$ | $F$ |
| $F$ | $F$ | $F$ | $F$ | $F$ |
| $T$ | $N$ | $F$ | $T$ | $B$ |
| $B$ | $F$ | $F$ | $B$ | $B$ |

| $\vee$ | $N$ | $F$ | $T$ | $B$ |
|---|---|---|---|---|
| $N$ | $N$ | $N$ | $T$ | $T$ |
| $F$ | $N$ | $F$ | $T$ | $B$ |
| $T$ | $T$ | $T$ | $T$ | $T$ |
| $B$ | $T$ | $B$ | $T$ | $B$ |

$s(A \wedge B) = s(A) \wedge s(B)$
$s(A \vee B) = s(A) \vee s(B)$
$s(\sim A) = \sim s(A)$

Further, Belnap defined an entailment relation $\rightarrow$ as follows:

$A \rightarrow B \Leftrightarrow s(A) \leq s(B)$

for all setups $s$.
The entailment relation $\rightarrow$ can be axiomatized as follows:

$(A_1 \wedge ... \wedge A_m) \rightarrow (B_1 \vee ... \vee B_n)$ ($A_i$ shares some $B_j$)
$(A \vee B) \rightarrow C \leftrightarrow (A \rightarrow C)$ and $(B \rightarrow C)$
$A \rightarrow B \Leftrightarrow \sim B \rightarrow \sim A$
$A \vee B \leftrightarrow B \vee A,\ A \wedge B \leftrightarrow B \wedge A$
$A \vee (B \vee C) \leftrightarrow (A \vee B) \vee C$
$A \wedge (B \wedge C) \leftrightarrow (A \wedge B) \wedge C$
$A \wedge (B \vee C) \leftrightarrow (A \wedge B) \vee (A \wedge C)$
$A \vee (B \wedge C) \leftrightarrow (A \vee B) \wedge (A \vee C)$
$(B \vee C) \wedge A \leftrightarrow (B \wedge A) \vee (C \wedge A)$
$(B \wedge C) \vee A \leftrightarrow (B \vee A) \wedge (C \vee A)$
$\sim\sim A \leftrightarrow A$
$\sim (A \wedge B) \leftrightarrow \sim A \vee \sim B, \quad \sim (A \vee B) \leftrightarrow \sim A \wedge \sim B$
$A \rightarrow B, B \rightarrow C \Leftrightarrow A \rightarrow C$
$A \leftrightarrow B, B \leftrightarrow C \Leftrightarrow A \leftrightarrow C$
$A \rightarrow B \Leftrightarrow A \leftrightarrow (A \wedge B) \Leftrightarrow (A \vee B) \leftrightarrow B$

Note here that $(A \wedge \sim A) \rightarrow B$ and $A \rightarrow (B \vee \sim B)$ cannot be derived in this axiomatization. It can be shown that the logic given above is closely related to the so-called *relevant logic* of Anderson and Belnap in [9]. In fact, Belnap's four-valued logic is equivalent to the system of *tautological entailment*.

*Infinite-valued logic* is a many-valued logic having infinite truth-values in [0, 1]. *fuzzy logic* and *probabilistic logic* belong to this family. Lukasiewicz introduced infinite-valued logic $\mathbf{L}_\infty$ in 1930; see Lukasiewicz [10]. Its truth-value tables can be generated by the following matrix:

$$
\begin{aligned}
|\sim A| &= 1 - |A| \\
|A \vee B| &= \max(|A|, |B|) \\
|A \wedge B| &= \min(|A|, |B|) \\
|A \to B| &= 1 \quad (|A| \leq |B|) \\
&= 1 - |A| + |B| \quad (|A| > |B|)
\end{aligned}
$$

A Hilbert system for $\mathbf{L}_\infty$ is as follows:

**Lukasiewicz's Infinite-Valued logic $\mathbf{L}_\infty$**
**Axioms**
(IL1) $A \to (B \to A)$
(IL2) $(A \to B) \to ((B \to C) \to (A \to C))$
(IL3) $((A \to B) \to B) \to ((B \to A) \to A)$
(IL4) $(\sim A \to \sim B) \to (B \to A)$
(IL5) $((A \to B) \to (B \to A)) \to (B \to A)$

**Rules of Inference**

(MP) $\vdash A, \vdash A \to B \Rightarrow \vdash B$

Since (IL5) derived from other axioms, it can be deleted.

It is known that $\mathbf{L}_\infty$ was used as the basis of *fuzzy logic* based on *fuzzy set* due to Zadeh [11]. Fuzzy logic is a logic of vagueness and is found in many applications. Since the 1990s, a lot of important work has been done for foundations for fuzzy logic.

Fitting [12, 13] studied *bilattice*, which is the lattice **4** with two kinds of orderings, in connection with the semantics of logic programs. Bilattices introduce non-standard logical connectives.

A *bilattice* was originally introduced by Ginsberg [14, 15] for the foundations of reasoning in AI, which has two kinds of orderings, i.e., truth ordering and knowledge ordering.

Later, it was extensively studied by Fitting in the context of logic programming in [16] and of theory of truth in [12]. In fact, bilattice-based logics can handle both incomplete and inconsistent information.

A *pre-bilattice* is a structure $\mathcal{B} = \langle B, \leq_t, \leq_k \rangle$, where $B$ denotes a non-empty set and $\leq_t$ and $\leq_k$ are partial orderings on $B$. The ordering $\leq_k$ is thought of as ranking "degree of information (or knowledge)". The bottom in $\leq_k$ is denoted by $\perp$ and the top by $\top$. If $x <_k y$, $y$ gives us at least as much information as $x$ (and possibly more).

The ordering $\leq_t$ is an ordering on the "degree of truth". The bottom in $\leq_t$ is denoted by *false* and the top by *true*. A bilattice can be obtained by adding certain assumptions for connections for two orderings.

One of the most well-known bilattices is the bilattice *FOUR* as depicted as Fig. 3.3. The billatice *FOUR* can be interpreted a combination of Belnap's lattices **A4** and **L4**.

The bilattice *FOUR* can be seen as Belnap's lattice *FOUR* with two kinds of orderings. Thus, we can think of the left-right direction as characterizing the ordering $\leq_t$: a move to the right is an increase in truth.

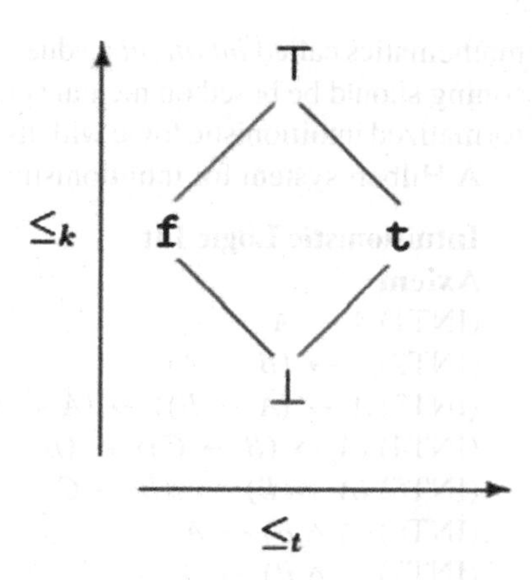


**Fig. 3.3** The bilattice *FOUR*

The meet operation $\wedge$ for $\leq_t$ is then characterized by: $x \wedge y$ is rightmost thing that is of left both $x$ and $y$. The join operation $\vee$ is dual to this. In a similar way, the up-down direction characterizes $\leq_k$: a move up is an increase in information. $x \otimes y$ is the uppermost thing below both $x$ and $y$, and $\oplus$ is its dual.

Fitting [16] gave a semantics for logic programming using bilattices. Kifer and Subrahmanian [17] interpreted Fitting's semantics within generalized annotated logics *GAL*.

Fitting [12] tried to generalize Kripke's [18] *theory of truth*, which is based on Kleene's strong three-valued logic, in a four-valued setting based on the bilattice *FOUR*.

A bilattice has a negation operation $\neg$ if there is a mapping $\neg$ that reverse $\leq_t$, leaves unchanged $\leq_k$ and $\neg\neg x = x$. Likewise a bilattice has a *conflation* if there is a mapping $-$ that reverse $\leq_k$, leaves unchanged $\leq_t$. and $- - x = x$. If a bilattice has both operations, they *commute* if $-\neg x = \neg - x$ for all $x$.

In the bilattice *FOUR*, there is a negation operator under which $\neg t = f$, $\neg f = t$, and $\bot$ and $\top$ are left unchanged. There is also a conflation under which $-\bot = \top$, $-\top = \bot$ and $t$ and $f$ are left unchanged. And negation and conflation commute. In any bilattice, if a negation or conflation exists then the extreme elements $\bot$, $\top$, $f$ and $t$ will behave as in *FOUR*.

Bilattice logics are theoretically elegant in that we can obtain several algebraic constructions, and are also suitable for reasoning about incomplete and inconsistent information. Arieli and Avron [19, 20] studied reasoning with bilattices. Thus, bilattice logics have many applications in AI as well as philosophy.

## 3.3 Intuitionistic Logic

*Intuitionistic logic* is a rival to classical logic in that it rejects the *law of excluded middle*, i.e., $A \vee \neg A$ in classical logic. Intuitionistic logic is a logic for philosophy of mathematics called *intuitionism* due to Brouwer who claimed that mathematical reasoning should be based on mental activity. Based on Brouwer's philosophy, Heyting formalized intuitionistic logic with its Hilbert style axiomatization; see Heyting [21].

A Hilbert system for intuitionistic logic **Int** is formalized as follows:

**Intuitionistic Logic Int**
**Axiom**
(INT1) $A \rightarrow A$
(INT2) $A \rightarrow (B \rightarrow A)$
(INT3) $A \rightarrow (A \rightarrow B)) \rightarrow (A \rightarrow B)$
(INT4) $(A \rightarrow (B \rightarrow C)) \rightarrow (B \rightarrow (A \rightarrow C))$
(INT5) $(A \rightarrow B) \rightarrow ((B \rightarrow C) \rightarrow (A \rightarrow C))$
(INT6) $(A \wedge B) \rightarrow A$
(INT7) $(A \wedge B) \rightarrow B$
(INT8) $(A \rightarrow B) \rightarrow ((A \rightarrow C) \rightarrow (A \rightarrow (B \wedge C)))$
(INT9) $A \rightarrow (A \vee B)$
(INT10) $B \rightarrow (A \vee B)$
(INT11) $(A \rightarrow C) \rightarrow ((B \rightarrow C) \rightarrow ((A \vee B) \rightarrow C))$
(INT12) $(A \rightarrow B) \rightarrow (A \rightarrow \neg B) \rightarrow \neg A)$
(INT13) $\neg A \rightarrow (A \rightarrow B)$
**Rules of Inference**

(MP) $\vdash A, \vdash A \rightarrow B \Rightarrow \vdash B$

The logical symbols used here are the same as the ones in classical logic. As in classical logic, intuitionistic negation $\neg A$ can be defined as $A \rightarrow false$. $\vdash_{\mathbf{INT}}$ is also used for provability.

If we add the law of excluded middle (LEM) or the *law of double negation* (LDN) to **INT**, we can get classical logic **CPC**.
(LEM) $A \vee \neg A$
(LDN) $\neg\neg A \rightarrow A$

A semantics of intuitionistic logic is also non-truth-functional. Kripke also developed a semantics for **INT**; see Kripke [22] and Fitting [13]. A Kripke semantics is similar to the one for modal logic **S4** due to the connection that **INT** can be embedded into **S4**.

A *Kripke model* for **INT** is defined as a triple $M = \langle W, R, V \rangle$, where

(1) $W$ is a non-empty set of possible worlds,
(2) $R$ is binary relation, which is reflexive and transitive, on $W$,
(3) $V$ is a valuation function which maps every propositional variable $p$ to a subset of $W$ satisfying $\forall w^*(w \in V(p) \Rightarrow w^* \in V(p))$.

Here, $\forall w^*$ abbreviates $\forall w^* \in W$ such that $wRw^*$.

We define a forcing relation $\models$ for any propositional variable $p$ and any $w \in W$ as follows:

$w \models p \Leftrightarrow w \in V(p)$

Here, $w \models p$ to mean that a formula $p$ is true at a world $w$. Then, $\models$ can be extended for any formula $A, B$ as follows:

$w \not\models \mathit{false}$
$w \models \neg A \Leftrightarrow \forall w^*(w^* \not\models A)$
$w \models A \wedge B \Leftrightarrow w \models A$ and $w \models B$
$w \models A \vee B \Leftrightarrow w \models A$ or $w \models B$
$w \models A \rightarrow B \Leftrightarrow \forall w^*(w^* \models A \Rightarrow w^* \models B)$

We say that $A$ is *valid*, written $\models_{\mathbf{INT}} A$, just in case $w \models A$ for every world $w$ and every model $M$.

Note that the monotonicity of $V$ holds for any formula. The striking feature of Kripke model for **INT** lies in the fact that both implication and negation are interpreted intensionally.

The completeness of **INT** holds.

**Theorem 3.2** $\vdash_{\mathbf{INT}} A \Leftrightarrow \models_{\mathbf{INT}} A$.

It is observed that we can provide an algebraic semantics for **INT** by Heyting algebras. The reader is invited to consult Fitting [16] for details on intuitionistic logic.

Intuitionistic logic was mainly studied from logical perspectives, but it receives special attention in computer science. Intuitionistic logic and its extensions, i.e., *intermediate logic* offer foundations for rough set theory, in particular, rough set logics.

The logics which are intermediate between intuitionistic logic and classical logic are called *intermediate logics* or *superintuitionistic logics*.

Many intermediate logics have been proposed in the literature. We here introduce some interesting ones. The *logic of the weak excluded middle* or Jankov's logic, denoted **KC** (or **LQ**), extends **INT** with $\neg\neg A \vee \neg A$; see Akama [23].

*Gödel-Dummett logic*, denoted **LC**, is an extension of **INT** with $(A \rightarrow B) \vee (B \rightarrow A)$; see Dummett [24].

*Kreisel-Putnam logic*, denoted **KP**, extends **INT** with $(\neg A \rightarrow (B \vee C)) \rightarrow ((\neg A \rightarrow B) \vee (\neg A \rightarrow C))$; see Krisel and Putnam [25].

Several intermediate predicate logics have also proposed, but we omit their review here. Observe that some intermediate logics are used as rough set logics; see Akama et al. [26, 27].

In intuitionistic logic, negation is not constructive, and it is possible to introduce strong negation into it. Nelson [28] proposed a *constructive logic with strong negation* as an alternative to intuitionistic logic, in which *strong negation* (or constructible negation) is introduced to improve some weaknesses of intuitionistic negation.

Constructive logic **N** extends positive intuitionistic logic $\mathbf{Int}^+$ with the following axioms for *strong negation* $\sim$:

(N1) $(A \wedge \sim A) \rightarrow B$
(N2) $\sim\sim A \leftrightarrow A$

(N3) $\sim (A \rightarrow B) \leftrightarrow (A \wedge \sim B)$
(N4) $\sim (A \wedge B) \leftrightarrow (\sim A \vee \sim B)$
(N5) $\sim (A \vee B) \leftrightarrow (\sim A \wedge \sim B)$

In **N**, intuitionistic negation $\neg$ can be defined by means of strong negation and implication. We can introduce it into **N** by one of the following:

$\neg A \leftrightarrow A \rightarrow (B \wedge \sim B)$
$\neg A \leftrightarrow A \rightarrow \sim A.$

If we delete (N1) from **N**, we can obtain a paraconsistent constructive logic $\mathbf{N}^-$ of Almukdad and Nelson [29]. Akama [30–35] extensively studied Nelson's constructive logics with strong negation including proof and model theory; also see Wansing [36].

In 1959, Nelson [37] developed a constructive logic **S** which lacks contraction $(A \rightarrow (A \rightarrow B)) \rightarrow (A \rightarrow B)$ and discussed its aspects as a paraconsistent logic. Akama [34] gave a detailed presentation of Nelson's paraconsistent constructive logics.

Semantics for **N** can be given by Kripke models or Nelson algebras. A Kripke model for **N** is a tuple $\langle W, R, V_P, V_N \rangle$, where $W$ is a set of possible worlds, $R$ is binary relation, which is reflexive and transitive, on $W$, and $V_P$ and $V_N$ are functions, each of which maps every propositional variable $p$ to a subset of $W$ satisfying:

(1) $V_P(p) \cap V_N(p) = \emptyset$,
(2) $\forall w^*(w \in V_P(p) \Rightarrow w^* \in V_P(p))$,
(3) $\forall w^*(w \in V_N(p) \Rightarrow w^* \in V_N(p))$.

We define two forcing relations $\models_P$ and $\models_N$ for any propositional variable $p$ and any $w \in W$ as follows:

$w \models_P p \Leftrightarrow w \in V_P(p)$
$w \models_N p \Leftrightarrow w \in V_N(p)$

Here, $w \models_P p$ to mean that a formula $p$ is true at a world $w$ and $w \models_N p$ to mean that a formula $p$ is false at a world $w$, respectively. Then, $\models_P$ and $\models_N$ can be extended for any formula $A$, $B$ as follows:

$w \models_P \sim A \Leftrightarrow w \models_N A$
$w \models_P A \wedge B \Leftrightarrow w \models_P A$ and $w \models_P B$
$w \models_P A \vee B \Leftrightarrow w \models_P A$ or $w \models_P B$
$w \models_P A \rightarrow B \Leftrightarrow \forall w^*(w^* \models_P A \Rightarrow w^* \models_P B)$
$w \models_N \sim A \Leftrightarrow w \models_P A$
$w \models_N A \wedge B \Leftrightarrow w \models_N A$ or $w \models_N B$
$w \models_N A \vee B \Leftrightarrow w \models_N A$ and $w \models_N B$
$w \models_N A \rightarrow B \Leftrightarrow w \models_P A$ and $w \models_N B$

We say that $A$ is *valid*, written $\models_{\mathbf{N}} A$, just in case $w \models_P A$ for every world $w$ and every model $M$.

Note that the monotonicity of $V_P$ and $V_N$ holds for any formula. The Kripke model for **N** is regarded as an extension of intuitionistic Kripke model in which both truth and falsity are interpreted intuitionistically. If the condtition $V_P(p) \cap V_N(p) = \emptyset$ is dropped, we can obtain a Kripke model for $\mathbf{N}^-$.

We have a completeness result of **N** (also $\mathbf{N}^-$).

**Theorem 3.3** $\vdash_{\mathbf{N}} A \Leftrightarrow \models_{\mathbf{N}} A$.

Algebraic semantics for constructive logic with strong negation has been studied using the so-called *Nelson algebras*; see Rasiowa [38]. Nelson algebras can serve as another basis for rough set theory, as will be discussed in Chap. 4.

## 3.4 Paraconsistent Logic

*Paraconsistent logic* is a logical system for inconsistent but non-trivial formal theories. It is classified as non-classical logic. Paraconsistent logic has many applications and it can serve as a foundation for engineering because some engineering problems must solve inconsistent information. However standard classical logic cannot tolerate it. In this regard, paraconsistent logic is promising.

Here, we give a quick review of paraconsistent logic, and it is helpful to the reader. Let $T$ be a theory whose underlying logic is $L$. $T$ is called *inconsistent* when it contains theorems of the form $A$ and $\neg A$ (the negation of $A$), i.e.,

$T \vdash_L A$ and $T \vdash_L \neg A$

where $\vdash_L$ denotes the provability relation in $L$. If $T$ is not inconsistent, it is called *consistent*.

$T$ is said to be *trivial*, if all formulas of the language are also theorems of $T$. Otherwise, $T$ is called *non-trivial*. Then, for trivial theory $T$, $T \vdash_L B$ for any formula $B$. Note that trivial theory is not interesting since every formula is provable.

If $L$ is classical logic (or one of several others, such as intuitionistic logic), the notions of inconsistency and triviality agree in the sense that $T$ is inconsistent iff $T$ is trivial. So, in trivial theories the extensions of the concepts of formula and theorem coincide.

A *paraconsistent logic* is a logic that can be used as the basis for inconsistent but non-trivial theories. In this regard, paraconsistent theories do not satisfy the *principle of non-contradiction*, i.e., $\neg(A \wedge \neg A)$.

Similarly, we can define the notion of paracomplete theory, namely $T$ is called *paracomplete* when neither $A$ nor $\neg A$ is a theorem. In other words,

$T \nvdash_L A$ and $T \nvdash_L \neg A$

hold in paracomplete theory. If $T$ is not paracomplete, $T$ is *complete*, i.e.,

$T \vdash_L A$ or $T \vdash_L \neg A$

holds. A *paracomplete logic* is a logic for paracomplete theory, in which the *principle of excluded middle*, i.e., $A \vee \neg A$ fails. In this sense, intuitionistic logic is one of the paracomplete logics.

Finally, a logic which is simultaneously paraconsistent and paracomplete is called *non-alethic logic*. Classical logic is a consistent and complete logic.

There are several systems of paraconsistent logic, which have been developed from different perspective. We here review the following three systems, since they are considered major ones.

- Discursive logic
- C-systems
- relevant (relevance) logic.

*Discursive logic*, also known as discussive logic, was proposed by Jaśkowski [39, 40], which is regarded as a non-adjunctive approach. *Adjunction* is a rule of inference of the form: from $\vdash A$ and $\vdash B$ to $\vdash A \wedge B$. Discursive logic can avoid explosion by prohibiting adjunction.

It was a formal system $J$ satisfying the conditions: (a) from two contradictory propositions, it should not be possible to deduce any proposition; (b) most of the classical theses compatible with (a) should be valid; (c) $J$ should have an intuitive interpretation.

Such a calculus has, among others, the following intuitive properties remarked by Jaśkowski himself: suppose that one desires to systematize in only one deductive system all theses defended in a discussion. In general, the participants do not confer the same meaning to some of the symbols.

One would have then as theses of a deductive system that formalize such a discussion, an assertion and its negation, so both are "true" since it has a variation in the sense given to the symbols. It is thus possible to regard discursive logic as one of the paraconsistent logics.

Jaśkowski's $D_2$ contains propositional formulas built from logical symbols of classical logic. In addition, the possibility operator $\diamond$ in S5 is added. Based on the possibility operator, three discursive logical symbols can be defined as follows:

discursive implication: $A \rightarrow_d B =_{\text{def}} \diamond A \rightarrow B$
discursive conjunction: $A \wedge_d B =_{\text{def}} \diamond A \wedge B$
discursive equivalence: $A \leftrightarrow_d B =_{\text{def}} (A \rightarrow_d B) \wedge_d (B \rightarrow_d A)$

Additionally, we can define discursive negation $\neg_d A$ as $A \rightarrow_d false$. Jaśkowski's original formulation of $D_2$ in [40] used the logical symbols: $\rightarrow_d$, $\leftrightarrow_d$, $\vee$, $\wedge$, $\neg$, and he later defined $\wedge_d$ in [40].

The following axiomatization due to Kotas [41] has the following axioms and the rules of inference.

**Axioms**
(A1) $\Box(A \rightarrow (\neg A \rightarrow B))$
(A2) $\Box((A \rightarrow B) \rightarrow ((B \rightarrow C) \rightarrow (A \rightarrow C))$
(A3) $\Box((\neg A \rightarrow A) \rightarrow A)$
(A4) $\Box(\Box A \rightarrow A)$

(A5) $\Box(\Box(A \rightarrow B) \rightarrow (\Box A \rightarrow \Box B))$
(A6) $\Box(\neg\Box A \rightarrow \Box\neg\Box A)$

**Rules of Inference**
(R1) substitution rule
(R2) $\Box A, \Box(A \rightarrow B)/\Box B$
(R3) $\Box A/\Box\Box A$
(R4) $\Box A/A$
(R5) $\neg\Box\neg\Box A/A$

There are other axiomatizations of $D_2$, but we omit the details here. Discursive logics are considered weak as a paraconsistent logic, but they have some applications, e.g. logics for vagueness.

*C-systems* are paraconsistent logics due to da Costa which can be a basis for inconsistent but non-trivial theories; see da Costa [42]. The important feature of da Costa systems is to use novel interpretation, which is non-truth-functional, of negation avoiding triviality.

Here, we review C-system $C_1$ due to da Costa [42]. The language of $C_1$ is based on the logical symbols: $\wedge$, $\vee$, $\rightarrow$, and $\neg$. $\leftrightarrow$ is defined as usual. In addition, a formula $A^\circ$, which is read "$A$ is well-behaved", is shorthand for $\neg(A \wedge \neg A)$.

The basic ideas of $C_1$ contain the following: (1) most valid formulas in the classical logic hold, (2) the law of non-contradiction $\neg(A \wedge \neg A)$ should not be valid, (3) from two contradictory formulas it should not be possible to deduce any formula.

The Hilbert system of $C_1$ extends the positive intuitionistic logic with the axioms for negation.

**da Costa's $C_1$**
**Axioms**
(DC1) $A \rightarrow (B \rightarrow A)$
(DC2) $(A \rightarrow B) \rightarrow (A \rightarrow (B \rightarrow C)) \rightarrow (A \rightarrow C))$
(DC3) $(A \wedge B) \rightarrow A$
(DC4) $(A \wedge B) \rightarrow B$
(DC5) $A \rightarrow (B \rightarrow (A \wedge B))$
(DC6) $A \rightarrow (A \vee B)$
(DC7) $B \rightarrow (A \vee B)$
(DC8) $(A \rightarrow C) \rightarrow ((B \rightarrow C) \rightarrow ((A \vee B) \rightarrow C))$
(DC9) $B^\circ \rightarrow ((A \rightarrow B) \rightarrow ((A \rightarrow \neg B) \rightarrow \neg A))$
(DC10) $(A^\circ \wedge B^\circ) \rightarrow (A \wedge B)^\circ \wedge (A \vee B)^\circ \wedge (A \rightarrow B)^\circ$
(DC11) $A \vee \neg A$
(DC12) $\neg\neg A \rightarrow A$

**Rules of Inference**
(MP) $\vdash A, \vdash A \rightarrow B \Rightarrow \vdash B$

Here, (DC1)–(DC8) are axioms of the positive intuitionistic logic. (DC9) and (DC10) play a role for the formalization of paraconsistency.

A semantics for $C_1$ can be given by a two-valued valuation; see da Costa and Alves [43]. We denote by $\mathscr{F}$ the set of formulas of $C_1$. A valuation is a mapping $v$ from $\mathscr{F}$ to $\{0, 1\}$ satisfying the following:

$v(A) = 0 \Rightarrow v(\neg A) = 1$
$v(\neg\neg A) = 1 \Rightarrow v(A) = 1$
$v(B^\circ) = v(A \rightarrow B) = v(A \rightarrow \neg B) = 1 \Rightarrow v(A) = 0$
$v(A \rightarrow B) = 1 \Leftrightarrow v(A) = 0$ or $v(B) = 1$
$v(A \wedge B) = 1 \Leftrightarrow v(A) = v(B) = 1$
$v(A \vee B) = 1 \Leftrightarrow v(A) = 1$ or $v(B) = 1$
$v(A^\circ) = v(B^\circ) = 1 \Rightarrow v((A \wedge B)^\circ) = v((A \vee B)^\circ) = v((A \rightarrow B)^\circ) = 1$

Note here that the interpretations of negation and double negation are not given by biconditional. A formula $A$ is *valid*, written $\models A$, if $v(A) = 1$ for every valuation $v$. Completeness holds for $C_1$. It can be shown that $C_1$ is complete for the above semantics.

Da Costa system $C_1$ can be extended to $C_n$ $(1 \leq n \leq \omega)$. Now, $A^{(1)}$ stands for $A^\circ$ and $A^{(n)}$ stands for $A^{(n-1)} \wedge (A^{(n-1)})^\circ$, $1 \leq n \leq \omega$.

Then, da Costa system $C_n$ $(1 \leq n \leq \omega)$ can be obtained by (DC1)–(DC8), (DC12), (DC13) and the following:

(DC9n) $B^{(n)} \rightarrow ((A \rightarrow B) \rightarrow ((A \rightarrow \neg B) \rightarrow \neg A))$
(DC10n) $(A^{(n)} \wedge B^{(n)}) \rightarrow (A \wedge B)^{(n)} \wedge (A \vee B)^{(n)} \wedge (A \rightarrow B)^{(n)}$

Note that the da Costa system $C_\omega$ has the axioms (DC1)–(DC8), (DC12) and (DC13). Later, da Costa investigated first-order and higher-order extensions of C-systems.

*Relevance logic*, also called *relevant logic*, is a family of logics based on the notion of relevance in conditionals. Historically, relevance logic was developed to avoid the *paradox of implications*; see Anderson and Belnap [9, 44].

Anderson and Belnap formalized a relevant logic **R** to realize a major motivation, in which they do not admit $A \rightarrow (B \rightarrow A)$. Later, various relevance logics have been proposed. Note that not all relevance logics are paraconsistent but some are considered important as paraconsistent logics.

Routley and Meyer proposed a basic relevant logic **B**, which is a minimal system having the so-called *Routley-Meyer semantics*. Thus, **B** is an important system and we review it below; see Routley et al. [45].

The language of **B** contains logical symbols: $\sim$, &, $\vee$ and $\rightarrow$ (relevant implication). A Hilbert system for **B** is as follows:

**Relevant Logic B**
**Axioms**
(BA1) $A \rightarrow A$
(BA2) $(A\&B) \rightarrow A$
(BA3) $(A\&B) \rightarrow B$
(BA4) $((A \rightarrow B)\&(A \rightarrow C)) \rightarrow (A \rightarrow (B\&C))$
(BA5) $A \rightarrow (A \vee B)$
(BA6) $B \rightarrow (A \vee B)$
(BA7) $(A \rightarrow C)\&(B \rightarrow C)) \rightarrow ((A \vee B) \rightarrow C)$
(BA8) $(A\&(B \vee C)) \rightarrow (A\&B) \vee C)$
(BA9) $\sim\sim A \rightarrow A$
**Rules of Inference**

(BR1) $\vdash A, \vdash A \rightarrow B \Rightarrow \vdash B$
(BR2) $\vdash A, \vdash B \Rightarrow \vdash A\&B$
(BR3) $\vdash A \rightarrow B, \vdash C \rightarrow D \Rightarrow \vdash (B \rightarrow C) \rightarrow (A \rightarrow D)$
(BR4) $\vdash A \rightarrow \sim B \Rightarrow \vdash B \rightarrow \sim A$

A Hilbert system for Anderson and Belnap's **R** is as follows:

**Relevance Logic R**
**Axioms**
(RA1) $A \rightarrow A$
(RA2) $(A \rightarrow B) \rightarrow ((C \rightarrow A) \rightarrow C \rightarrow B))$
(RA3) $(A \rightarrow (A \rightarrow B) \rightarrow (A \rightarrow B)$
(RA4) $(A \rightarrow (B \rightarrow C)) \rightarrow (B \rightarrow (A \rightarrow C)$
(RA5) $(A\&B) \rightarrow A$
(RA6) $(A\&B) \rightarrow B$
(RA7) $((A \rightarrow B)\&(A \rightarrow C)) \rightarrow (A \rightarrow (B\&C))$
(RA8) $A \rightarrow (A \vee B)$
(RA9) $B \rightarrow (A \vee B)$
(RA10) $((A \rightarrow C)\&(B \vee C)) \rightarrow ((A \vee B) \rightarrow C))$
(RA11) $(A\&(B \vee C)) \rightarrow ((A\&B) \vee C)$
(RA12) $(A \rightarrow \sim A) \rightarrow \sim A$
(RA13) $(A \rightarrow \sim B)) \rightarrow (B \rightarrow \sim A)$
(RA14) $\sim\sim A \rightarrow A$
**Rules of Inference**
(RR1) $\vdash A, \vdash A \rightarrow B \Rightarrow \vdash B$
(RR2) $\vdash A, \vdash B \Rightarrow \vdash A\&B$

Routley et al. considered some axioms of **R** are too strong and formalized as rules instead of axioms. Notice that **B** is a paraconsistent but **R** is not.

Next, we give a Routley-Meyer semantics for **B**. A *model structure* is a tuple $\mathcal{M} = \langle K, N, R, *, v \rangle$, where $K$ is a non-empty set of worlds, $N \subseteq K$, $R \subseteq K^3$ is a ternary relation on $K$, $*$ is a unary operation on $K$, and $v$ is a valuation function from a set of worlds and a set of propositional variables $\mathcal{P}$ to $\{0, 1\}$.

There are some restrictions on $\mathcal{M}$. $v$ satisfies the condition that $a \leq b$ and $v(a, p) = 1$ imply $v(b, p) = 1$ for any $a, b \in K$ and any $p \in \mathcal{P}$. $a \leq b$ is a pre-order relation defined by $\exists x(x \in N$ and $Rxab)$. The operation $*$ satisfies the condition $a^{**} = a$.

For any propositional variable $p$, the truth condition $\models$ is defined: $a \models p$ iff $v(a, p) = 1$. Here, $a \models p$ reads "$p$ is true at $a$". $\models$ can be extended for any formulas in the following way:

$a \models \sim A \Leftrightarrow a^* \not\models A$
$a \models A\&B \Leftrightarrow a \models A$ and $a \models B$
$a \models A \vee B \Leftrightarrow a \models A$ or $a \models B$
$a \models A \rightarrow B \Leftrightarrow \forall bc \in K(Rabc$ and $b \models A \Rightarrow c \models B)$

A formula $A$ is *true* at $a$ in $\mathcal{M}$ iff $a \models A$. $A$ is *valid*, written $\models A$, iff $A$ is true on all members of $N$ in all model structures.

Routley et al. provided the completeness theorem for **B** with respect to the above semantics using canonical models; see [45].

A model structure for **R** needs the following conditions.

$R0aa$
$Rabc \Rightarrow Rbac$
$R^2(ab)cd \Rightarrow R^2a(bc)d$
$Raaa$
$a^{**} = a$
$Rabc \Rightarrow Rac^*b^*$
$Rabc \Rightarrow (a' \leq a \Rightarrow Ra'bc)$

where $R^2abcd$ is shorthand for $\exists x(Raxd \text{ and } Rxcd)$. The completeness theorem for the Routley-Meyer semantics can be proved for **R**; see [9, 44].

The reader is advised to consult Anderson and Belnap [9], Anderson, Belnap and Dunn [44], and Routley et al. [45] for details. A more concise survey on the subject may be found in Dunn [46].

In the 1990s paraconsistent logics became one of the major topics in logic in connection with other areas, in particular, computer science. Below we review some of those systems of paraconsistent logics.

The modern history of paraconsistent logic started with Vasil'ev's *imaginary logic*. In 1910, Vasil'ev proposed an extension of Aristotle's syllogistic allowing the statement of the form $S$ is both $P$ and not-$P$; see Vasil'ev [47].

Thus, imaginary logic can be viewed as a paraconsistent logic. Unfortunately, little work has been done on focusing on its formalization from the viewpoint of modern logic. A survey of imaginary logic can be found in Arruda [48].

In 1954, Asenjo developed a calculus of antinomies in his dissertation; see Asenjo [49]. Asenjo's work was published before da Costa's work, but it seems that Asenjo's approach has been neglected. Asenjo's idea is to interpret the truth-value of *antinomy* as both true and false using Kleene's strong three-valued logic.

His proposed calculus is non-trivially inconsistent propositional logic, whose axiomatization can be obtained from Kleene's [6] axiomatization of classical propositional logic by deleting the axiom $(A \rightarrow B) \rightarrow ((A \rightarrow \neg B) \rightarrow \neg A)$.

In 1979, Priest [50] proposed a *logic of paradox*, denoted $LP$, to deal with the semantic paradox. The logic is of special importance to the area of paraconsistent logics. $LP$ can be semantically defined by Kleene's strong three-valued logic.

Priest re-interpreted the truth-value tables of Kleene's strong three-valued logic, namely read the third-truth value as both true and false ($B$) rather than neither true nor false ($I$), and assumed that ($T$) and ($B$) are designated values. The idea has already been considered in Asenjo [49] and Belnap [7, 8].

Consequently, ECQ: $A, \sim A \models B$ is invalid. Thus, $LP$ can be seen as a paraconsistent logic. Unfortunately, (material) implication in $LP$ does not satisfy *modus ponens*. It is, however, possible to introduce relevant implications as real implication into $LP$.

Priest developed a semantics for *LP* by means of a truth-value assignment relation rather than a truth-value assignment function. Let $\mathscr{P}$ be the set of propositional variables. Then, an evaluation $\eta$ is a subset of $\mathscr{P} \times \{0, 1\}$.

A proposition may only relate to 1 (true), it may only relate to 0 (false), it may relate to both 1 and 0 or it may relate to neither 1 nor 0. The evaluation is extended to a relation for all formulas as follows:

$\neg A\eta 1$ iff $A\eta 0$
$\neg A\eta 0$ iff $A\eta 1$
$A \wedge B\eta 1$ iff $A\eta 1$ and $B\eta 1$
$A \wedge B\eta 0$ iff $A\eta 0$ or $B\eta 0$
$A \vee B\eta 1$ iff $A\eta 1$ or $B\eta 1$
$A \vee B\eta 0$ iff $A\eta 0$ and $B\eta 0$

If we define validity in terms of truth preservation under all relational evaluations, then we obtain *first-degree entailment* which is a fragment of relevance logics.

Using *LP*, Priest advanced his research program to tackle various philosophical and logical issues; see Priest [51, 52] for details. For instance, in *LP*, the *liar sentence* can be interpreted as both true and false.

It is also observed that Priest promoted the philosophical view called *dialetheism* which claims that there are true contradictions. And dialetheism has been extensively discussed in philosophical logic by many people.

Since the beginning of the 1990s, Batens developed the so-called *adaptative logics* in Batens [53, 54]. These logics are considered as improvements of *dynamic dialectical logics* investigated in Batens [55]. *Inconsistency-adaptive logics* as developed by Batens [53] can serve as foundations for paraconsistent and non-monotonic logics.

Adaptive logics formalized classical logic as "dynamic logic". Here, "dynamic logic" is not the family of logics with the same name studied in computer science. A logic is *adaptive* iff it adapts itself to the specific premises to which it is applied. In this sense, adaptive logics can model the dynamics of human reasoning. There are two sorts of dynamics, i.e., *external dynamics* and *internal dynamics*.

The external dynamics is stated as follows. If new premises become available, then consequences derived from the earlier premise set may be withdrawn. In other words, the external dynamics results from the *non-monotonic* character of the consequence relations.

Let $\vdash$ be a consequence relation, $\Gamma$, $\Delta$ be sets of formulas, and $A$ be a formula. Then, the external dynamics is formally presented as: $\Gamma \vdash A$ but $\Gamma \cup \Delta \nvdash A$ for some $\Gamma$, $\Delta$ and $A$. In fact, the external dynamics is closely related to the notion of *non-monotonic reasoning* in AI.

The internal dynamics is very different from the external one. Even if the premise set is constant, certain formulas are considered as derived at some stage of the reasoning process, but are considered as not derived at a later stage. For any consequence relation, insight in the premises is gained by deriving consequences from them.

In the absence of a positive test, this results in the internal dynamics. Namely, in the internal dynamics, reasoning has to adapt itself by withdrawing an application of the previously used inference rule, if we infer a contradiction at a later stage. Adaptive logics are logics based on the internal dynamics.

An Adaptive Logic *AL* can be characterized as a triple:

(i) A *lower limit logic* (LLL)
(ii) A set of *abnormalities*
(iii) An *adaptive strategy*.

The lower limit logic *LLL* is any monotonic logic, e.g., classical logic, which is the stable part of the adaptive logic. Thus, *LLL* is not subject to adaptation. The set of abnormalities $\Omega$ comprises the formulas that are presupposed to be false, unless and until proven otherwise.

In many adaptive logics, $\Omega$ is the set of formulas of the form $A \wedge \sim A$. An adaptive strategy specifies a strategy of the applications of inference rules based on the set of abnormalities.

If the lower limit logic *LLL* is extended with the requirement that no abnormality is logically possible, one obtains a monotonic logic, which is called the *upper limit logic ULL*. Semantically, an adequate semantics for the upper limit logic can be obtained by selecting that lower limit logic models that verify no abnormality.

The name "abnormality" refers to the upper limit logic. *ULL* requires premise sets to be normal, and 'explodes' abnormal premise sets (assigns them the trivial consequence set).

If the lower limit logic is classical logic *CL* and the set of abnormalities comprises formulas of the form $\exists A \wedge \exists \sim A$, then the upper limit logic obtained by adding to *CL* the axioms $\exists A \rightarrow \forall A$. If, as is the case for many inconsistency-adaptive logics, the lower limit logic is a paraconsistent logic *PL* which contains *CL*, and the set of abnormalities comprises the formulas of the form $\exists(A \wedge \sim A)$, then the upper limit logic is *CL*.

The adaptive logics interpret the set of premises 'as much as possible' in agreement with the upper limit logic; it avoids abnormalities 'in as far as the premises permit'.

Adaptive logics provide a new way of thinking of the formalization of paraconsistent logics in view of the dynamics of reasoning. Although inconsistency-adaptive logic is paraconsistent logic, applications of adaptive logics are not limited to paraconsistency. From a formal point of view, we can count adaptive logics as promising paraconsistent logics.

However, for applications, we may face several obstacles in automating reasoning in adaptive logics in that proofs in adaptive logics are dynamic with a certain adaptive strategy. Thus, the implementation is not easy, and we have to choose an appropriate adaptive strategy depending on applications.

Carnelli proposed the *Logics of Formal Inconsistency* (LFI), which are logical systems that treat consistency and inconsistency as mathematical objects; see Carnelli, Coniglio and Marcos [56]. One of the distinguishing features of these logics is that they can internalize the notions of consistency and inconsistency at the object-level.

And many paraconsistent logics including da Costa's C-systems can be interpreted as the subclass of LFIs. Therefore, we can regard LFIs as a general framework for paraconsistent logics.

A Logic of Formal Inconsistency, which extends classical logic $C$ with the consistency operator $\circ$, is defined as any explosive paraconsistent logic, namely iff the classical consequence relation $\vdash$ satisfies the following two conditions:

(a) $\exists\Gamma\exists A\exists B(\Gamma, A, \neg A \nvdash B)$
(b) $\forall\Gamma\forall A\forall B(\Gamma, \circ A, A, \neg A \vdash B)$.

Here, $\Gamma$ denotes a set of formulas and $A$, $B$ are formulas. With the help of $\circ$, we can express both consistency and inconsistency in the object-language. Therefore, LFIs are general enough to classify many paraconsistent logics.

For example, da Costa's $C_1$ is shown to be an LFI. For every formula $A$, let $\circ A$ be an abbreviation of the formula $\neg(A \wedge \neg A)$. Then, the logic $C_1$ is an LFI such that $\circ(p) = \{\circ p\} = \{\neg\neg(p \wedge \neg p\}$ whose axiomatization as an LFI contains the positive fragment of classical logic with the axiom $\neg\neg A \rightarrow A$, and some axioms for $\circ$.

(bc1) $\circ A \rightarrow (A \rightarrow (\neg A \rightarrow B))$
(ca1) $(\circ A \wedge \circ B) \rightarrow \circ(A \wedge B)$
(ca2) $(\circ A \wedge \circ B) \rightarrow \circ(A \vee B)$
(ca3) $(\circ A \wedge \circ B) \rightarrow \circ(A \rightarrow B)$

In addition, we can define classical negation $\sim$ by $\sim A =_{def} \neg A \wedge \circ A$. If needed, the inconsistency operator $\bullet$ is introduced by definition:
$\bullet A =_{\text{def}} \neg \circ A$.

Carnielli, Coniglio, and Marcos [56] showed classifications of existing logical systems. For example, classical logic is not an LFI, and Jáskowski's $D_2$ is an LFI. They also introduced a basic system of LFI, called LFI1, with a semantics and axiomatization.

We can thus see that the Logics of Formal Inconsistency are very interesting from a logical point of view in that they can serve as a theoretical framework for existing paraconsistent logics. In addition, there are tableau systems for LFIs; see Carnielli and Marcos [57], and they can be properly applied to various areas including computer science and AI.

*Annotated logic* is a logic for paraconsistent logic programming; see Subrahmanian [58, 59]. It is also regarded as one of the attractive paraconsistent logics; see da Costa et al. [60, 61] and Abe [62]. Note that annotated logic has many applications for several areas including engineering.

Annotated logics were introduced by Subrahmanian to provide a foundation for *paraconsistent logic programming*; see Subrahmanian [58] and Blair and Subrahmanian [59]. Paraconsistent logic programming can be seen as an extension of logic programming based on classical logic.

Now, we formally introduce annotated logics. We denote by the language of propositional annotated logics $P\tau$ by $L$. Annotated logics are based on some arbitrary fixed finite lattice called a *lattice of truth-values*, denoted by $\tau = \langle | \tau |, \leq, \sim \rangle$, which is the complete lattice with the ordering $\leq$ and the operator $\sim: | \tau | \rightarrow | \tau |$.

Here, $\sim$ gives the "meaning" of atomic-level negation of $P\tau$. We also assume that $\top$ is the top element and $\bot$ is the bottom element, respectively. In addition, we use two lattice-theoretic operations: $\vee$ for the least upper bound and $\wedge$ for the greatest lower bound.[1]

**Definition 3.1** (*Symbols*) The symbols of $P\tau$ are defined as follows:

1. Propositional symbols: $p, q, \ldots$ (possibly with subscript)
2. Annotated constants: $\mu, \lambda, \ldots \in |\tau|$
3. Logical connectives: $\wedge$ (conjunction), $\vee$ (disjunction), $\rightarrow$ (implication), and $\neg$ (negation)
4. Parentheses: (and)

**Definition 3.2** (*Formulas*) Formulas are defined as follows:

1. If $p$ is a propositional symbol and $\mu \in |\tau|$ is an annotated constant, then $p_\mu$ is a formula called an *annotated atom*.
2. If $F$ is a formula, then $\neg F$ is a formula.
3. If $F$ and $G$ are formulas, then $F \wedge G, F \vee G, F \rightarrow G$ are formulas.
4. If $p$ is a propositional symbol and $\mu \in |\tau|$ is an annotated constant, then a formula of the form $\neg^k p_\mu$ $(k \geq 0)$ is called a *hyper-literal*. A formula which is not a hyper-literal is called a *complex formula*.

Here, some remarks are in order. The annotation is attached only at the atomic level. An annotated atom of the form $p_\mu$ can be read "it is believed that $p$'s truth-value is at least $\mu$". In this sense, annotated logics incorporate the feature of many-valued logics.

A hyper-literal is special kind of formula in annotated logics. In the hyper-literal of the form $\neg^k p_\mu$, $\neg^k$ denotes the $k$'s repetition of $\neg$. More formally, if $A$ is an annotated atom, then $\neg^0 A$ is $A$, $\neg^1 A$ is $\neg A$, and $\neg^k A$ is $\neg(\neg^{k-1} A)$. The convention is also use for $\sim$.

Next, we define some abbreviations.

**Definition 3.3** Let $A$ and $B$ be formulas. Then, we put:
$A \leftrightarrow B =_{\text{def}} (A \rightarrow B) \wedge (B \rightarrow A)$
$\neg_* A =_{\text{def}} A \rightarrow (A \rightarrow A) \wedge \neg(A \rightarrow A)$
Here, $\leftrightarrow$ is called the *equivalence* and $\neg_*$ *strong negation*, respectively.

Observe that strong negation in annotated logics behaves classically in that it has all the properties of classical negation.

We turn to a semantics for $P\tau$. We here describe a *model-theoretic semantics* for $P\tau$. Let $\mathbf{P}$ is the set of propositional variables. An *interpretation* $I$ is a function $I : \mathbf{P} \rightarrow \tau$. To each interpretation $I$, we associate a *valuation* $v_I : \mathbf{F} \rightarrow \mathbf{2}$, where $\mathbf{F}$ is a set of all formulas and $\mathbf{2} = \{0, 1\}$ is the set of truth-values. Henceforth, the subscript is suppressed when the context is clear.

[1] We employ the same symbols for lattice-theoretical operations as the corresponding logical connectives.

**Definition 3.4** (*Valuation*) A valuation $v$ is defined as follows:
If $p_\lambda$ is an annotated atom, then
$v(p_\lambda) = 1$ iff $I(p) \geq \lambda$,
$v(p_\lambda) = 0$ otherwise,
$v(\neg^k p_\lambda) = v(\neg^{k-1} p_{\sim\lambda})$, where $k \geq 1$.
If $A$ and $B$ are formulas, then
$v(A \wedge B) = 1$ iff $v(A) = v(B) = 1$,
$v(A \vee B) = 0$ iff $v(A) = v(B) = 0$,
$v(A \rightarrow B) = 0$ iff $v(A) = 1$ and $v(B) = 0$.
If $A$ is a complex formula, then
$v(\neg A) = 1 - v(A)$.

Say that the valuation $v$ *satisfies* the formula $A$ if $v(A) = 1$ and that $v$ *falsifies* $A$ if $v(A) = 0$. For the valuation $v$, we can obtain the following lemmas.

**Lemma 3.1** *Let p be a propositional variable and* $\mu \in |\tau|$ $(k \geq 0)$, *then we have:* $v(\neg^k p_\mu) = v(p_{\sim^k \mu})$.

**Lemma 3.2** *Let p be a propositional variable, then we have:*
$v(p_\perp) = 1$

**Lemma 3.3** *For any complex formula A and B and any formula F, the valuation v satisfies the following:*

1. $v(A \leftrightarrow B) = 1$ iff $v(A) = v(B)$
2. $v((A \rightarrow A) \wedge \neg(A \rightarrow A)) = 0$
3. $v(\neg_* A) = 1 - v(A)$
4. $v(\neg F \leftrightarrow \neg_* F) = 1$.

We here define the notion of semantic consequence relation denoted by $\models$. Let $\Gamma$ be a set of formulas and $F$ be a formula. Then, $F$ is a *semantic consequence* of $\Gamma$, written $\Gamma \models F$, iff for every $v$ such that $v(A) = 1$ for each $A \in \Gamma$, it is the case that $v(F) = 1$.

If $v(A) = 1$ for each $A \in \Gamma$, then $v$ is called a *model* of $\Gamma$. If $\Gamma$ is empty, then $\Gamma \models F$ is simply written as $\models F$ to mean that $F$ is *valid*.

**Lemma 3.4** *Let p be a propositional variable and* $\mu, \lambda \in |\tau|$. *Then, we have:*

1. $\models p_\perp$
2. $\models p_\mu \rightarrow p_\lambda, \ \mu \geq \lambda$
3. $\models \neg^k p_\mu \leftrightarrow p_{\sim^k \mu}, \ k \geq 0$.

The consequence relation $\models$ satisfies the next property.

**Lemma 3.5** *Let A, B be formulas. Then, if* $\models A$ *and* $\models A \rightarrow B$ *then* $\models B$.

**Lemma 3.6** *Let F be a formula and p a propositional variable.* $(\mu_i)_{i \in J}$ *be an annotated constant, where J is an indexed set. Then, if* $\models F \rightarrow p_\mu$, *then* $F \rightarrow p_{\mu_i}$, *where* $\mu = \bigvee \mu_i$.

As a corollary to Lemma 3.6, we can obtain the following lemma.

**Lemma 3.7** $\models p_{\lambda_1} \wedge p_{\lambda_2} \wedge \ldots \wedge p_{\lambda_m} \rightarrow p_\lambda$, *where* $\lambda = \bigvee_{i=1}^{m} \lambda_i$.

Next, we discuss some results related to paraconsistency and paracompleteness.

**Definition 3.5** (*Complementary property*) A truth-value $\mu \in \tau$ has the *complementary property* if there is a $\lambda$ such that $\lambda \leq \mu$ and $\sim \lambda \leq \mu$. A set $\tau' \subseteq \tau$ has the *complementary property* iff there is some $\mu \in \tau'$ such that $\mu$ has the complementary property.

**Definition 3.6** (*Range*) Suppose $I$ is an interpretation of the language $L$. The *range* of $I$, denoted *range*$(I)$, is defined to be *range*$(I) = \{\mu \mid (\exists A \in B_L) I(A) = \mu\}$, where $B_L$ denotes the set of all ground atoms in $L$.

For $P\tau$, ground atoms correspond to propositional variables. If the range of the interpretation $I$ satisfies the complementary property, then the following theorem can be established.

**Theorem 3.4** *Let* $I$ *be an interpretation such that range*$(I)$ *has the complementary property. Then, there is a propositional variable* $p$ *and* $\mu \in |\tau|$ *such that* $v(p_\mu) = v(\neg p_\mu) = 1$.

Theorem 3.1 states that there is a case in which for some propositional variable it is both true and false, i.e., inconsistent. The fact is closely tied with the notion of paraconsistency.

**Definition 3.7** ($\neg$-*inconsistency*) We say that an interpretation $I$ is $\neg$-*inconsistent* iff there is a propositional variable $p$ and an annotated constant $\mu \in |\tau|$ such that $v(p_\mu) = v(\neg p_\mu) = 1$.

Therefore, $\neg$-inconsistency means that both $A$ and $\neg A$ are simultaneously true for some atomic $A$. Below, we formally define the concepts of non-triviality, paraconsistency and paracompleteness.

**Definition 3.8** (*Non-triviality*) We say that an interpretation $I$ is *non-trivial* iff there is a propositional variable $p$ and an annotated constant $\mu \in |\tau|$ such that $v(p_\mu) = 0$.

By Definition 3.8, we mean that not every atom is valid if an interpretation is non-trivial.

**Definition 3.9** (*Paraconsistency*) We say that a interpretation $I$ is *paraconsistent* iff it is both $\neg$-inconsistent and non-trivial. $P\tau$ is called *paraconsistent* iff there is an interpretation of $I$ of $P\tau$ such that $I$ is paraconsistent.

Definition 3.9 allows the case in which both $A$ an $\neg A$ are true, but some formula $B$ is false in some paraconsistent interpretation $I$.

**Definition 3.10** (*Paracompleteness*) We say that an interpretation $I$ is *paracomplete* iff there is a propositional variable $p$ and a annotated constant $\lambda \in |\tau|$ such that $v(p_\lambda) = v(\neg p_\lambda) = 0$. $P\tau$ is called *paracomplete* iff there is an interpretation $I$ of $P\tau$ such that $I$ is paracomplete.

From Definition 3.10, we can see that in the paracomplete interpretation $I$, both $A$ and $\neg A$ are false. We say that $P\tau$ is *non-alethic* iff it is both paraconsistent and paracomplete.

Intuitively speaking, paraconsistent logic can deal with inconsistent information and paracomplete logic can handle incomplete information.

This means that non-alethic logics like annotated logics can serve as logics for expressing both inconsistent and incomplete information. This is one of the starting points of our study of annotated logics.

As the following Theorems 3.2 and 3.3 indicate, paraconsistency and paracompleteness in $P\tau$ depend on the cardinality of $\tau$.

**Theorem 3.5** *$P\tau$ is paraconsistent iff card$(\tau) \geq 2$, where card$(\tau)$ denotes the cardinality (cardinal number) of the set $\tau$.*

**Theorem 3.6** *If card$(\tau) \geq 2$, then there are annotated systems $P\tau$ such that they are paracomplete.*

The above two theorems imply that to formalize a non-alethic logic based on annotated logics we need at least both the top and bottom elements of truth-values. The simplest lattice of truth-values is *FOUR* in Belnap [7, 8].

**Definition 3.11** (*Theory*) Given an interpretation $I$, we can define the theory $Th(I)$ associated with $I$ to be a set:
$Th(I) = Cn(\{p_\mu \mid p \in \mathbf{P} \text{ and } I(p) \geq \mu\})$.
Here, $Cn$ is the semantic consequence relation, i.e.,
$Cn(\Gamma) = \{F \mid F \in \mathbf{F} \text{ and } \Gamma \models F\}$.
Here, $\Gamma$ is a set of formulas.

$Th(I)$ can be extended for any set of formulas.

**Theorem 3.7** *An interpretation $I$ is $\neg$-inconsistent iff $Th(\Gamma)$ is $\neg$-inconsistent.*

**Theorem 3.8** *An interpretation $I$ is paraconsistent iff $Th(I)$ is paraconsistent.*

The next lemma states that the replacement of equivalent formulas within the scope of $\neg$ does not hold in $P\tau$ as in other paraconsistent logics.

**Lemma 3.8** *Let $A$ be any hyper-literal. Then, we have:*

1. $\models A \leftrightarrow ((A \rightarrow A) \rightarrow A)$
2. $\not\models \neg A \leftrightarrow \neg(((A \rightarrow A) \rightarrow A))$
3. $\models A \leftrightarrow (A \wedge A)$
4. $\not\models \neg A \leftrightarrow \neg(A \wedge A)$

5. $\models A \leftrightarrow (A \vee A)$
6. $\not\models \neg A \leftrightarrow \neg(A \vee A)$.

As obvious from the above proofs, (1), (3) and (5) hold for any formula $A$. But, (2), (4) and (6) cannot be generalized for any $A$.

By the next theorem, we can find the connection of $P\tau$ and the positive fragment of classical propositional logic $C$.

**Theorem 3.9** *If $F_1, ..., F_n$ are complex formulas and $K(A_1, ..., A_n)$ is a tautology of $C$, where $A_1, ..., A_n$ are the sole propositional variable occurring in the tautology, then $K(F_1, ..., F_n)$ is valid in $P\tau$. Here, $K(F_1, ..., F_n)$ is obtained by replacing each occurrence of $A_i$, $1 \leq i \leq n$, in $K$ by $F_i$.*

Next, we consider the properties of strong negation $\neg_*$.

**Theorem 3.10** *Let $A, B$ be any formulas. Then,*

1. $\models (A \to B) \to ((A \to \neg_* B) \to \neg_* A)$
2. $\models A \to (\neg_* A \to B)$
3. $\models A \vee \neg_* A$.

Theorem 3.10 tells us that strong negation has all the basic properties of classical negation. Namely, (1) is a principle of *reductio ad abusurdum*, (2) is the related principle of the law of non-contradiction, and (3) is the law of excluded middle. Note that $\neg$ does not satisfy these properties. It is also noticed that for any complex formula $A \models \neg A \leftrightarrow \neg_* A$ but that for any hyper-literal $Q \not\models \neg Q \leftrightarrow \neg_* Q$.

From these observations, $P\tau$ is a paraconsistent and paracomplete logic, but adding strong negation enables us to perform classical reasoning.

Next, we provide an axiomatization of $P\tau$ in the Hilbert style. There are many ways to axiomatize a logical system, one of which is the *Hilbert system*. Hilbert system can be defined by the set of *axioms* and *rules of inference*. Here, an axiom is a formula to be postulated as valid, and rules of inference specify how to prove a formula.

We are now ready to give a Hilbert style axiomatization of $P\tau$, called $\mathscr{A}\tau$. Let $A, B, C$ be arbitrary formulas, $F, G$ be complex formulas, $p$ be a propositional variable, and $\lambda, \mu, \lambda_i$ be annotated constant. Then, the postulates are as follows (cf. Abe [62]):

**Postulates for $\mathscr{A}\tau$**

($\to_1$) $(A \to (B \to A)$
($\to_2$) $(A \to (B \to C)) \to ((A \to B) \to (A \to C))$
($\to_3$) $((A \to B) \to A) \to A$
($\to_4$) $A, A \to B / B$
($\wedge_1$) $(A \wedge B) \to A$
($\wedge_2$) $(A \wedge B) \to B$
($\wedge_3$) $A \to (B \to (A \wedge B))$
($\vee_1$) $A \to (A \vee B)$
($\vee_2$) $B \to (A \vee B)$

$(\vee_3)$ $(A \to C) \to ((B \to C) \to ((A \vee B) \to C))$
$(\neg_1)$ $(F \to G) \to ((F \to \neg G) \to \neg F)$
$(\neg_2)$ $F \to (\neg F \to A)$
$(\neg_3)$ $F \vee \neg F$
$(\tau_1)$ $p_\perp$
$(\tau_2)$ $\neg^k p_\lambda \leftrightarrow \neg^{k-1} p_{\sim\lambda}$
$(\tau_3)$ $p_\lambda \to p_\mu$, where $\lambda \geq \mu$
$(\tau_4)$ $p_{\lambda_1} \wedge p_{\lambda_2} \wedge \ldots \wedge p_{\lambda_m} \to p_\lambda$, where $\lambda = \bigvee_{i=1}^{m} \lambda_i$

Here, except $(\to_4)$, these postulates are axioms. $(\to_4)$ is a rule of inferences called *modus ponens* (MP).

In da Costa, Subrahmanian and Vago [60], a different axiomatization is given, but it is essentially the same as ours. There, the postulates for implication are different. Namely, although $(\to_1)$ and $(\to_3)$ are the same (although the naming differs), the remaining axiom is:

$$(A \to B) \to ((A \to (B \to C)) \to (A \to C))$$

It is well known that there are many ways to axiomatize the implicational fragment of classical logic $C$. In the absence of negation, we need the so-called *Pierce's law* $(\to_3)$ for $C$.

In $(\neg_1)$, $(\neg_2)$, $(\neg_3)$, $F$ and $G$ are complex formulas. In general, without this restriction on $F$ and $G$, these are not sound rules due to the fact that they are not admitted in annotated logics.

Da Costa, Subrahmanian and Vago [60] fuses $(\tau_1)$ and $(\tau_2)$ as the single axiom in conjunctive form. But, we separate it in two axioms for our purposes. Also there is a difference in the final axiom. They present it for infinite lattices as
$A \to p_{\lambda_j}$ for every $j \in J$, then $A \to p_\lambda$, where $\lambda = \bigvee_{j \in J} \lambda_j$.
If $\tau$ is a finite lattice, this is equivalent to the form of $(\tau_2)$.

As usual, we can define a *syntactic consequence relation* in $P\tau$. Let $\Gamma$ be a set of formulas and $G$ be a formula. Then, $G$ is a syntactic consequence of $\Gamma$, written $\Gamma \vdash G$, iff there is a finite sequence of formulas $F_1, F_2, \ldots, F_n$, where $F_i$ belongs to $\Gamma$, or $F_i$ is an axiom $(1 \leq i \leq n)$, or $F_j$ is an immediate consequence of the previous two formulas by $(\to_4)$. This definition can extend for the transfinite case in which $n$ is an ordinal number. If $\Gamma = \emptyset$, i.e. $\vdash G$, $G$ is a *theorem* of $P\tau$.

Let $\Gamma$, $\Delta$ be sets of formulas and $A$, $B$ be formulas. Then, the consequence relation $\vdash$ satisfies the following conditions.

1. if $\Gamma \vdash A$ and $\Gamma \subset \Delta$ then $\Delta \vdash A$.
2. if $\Gamma \vdash A$ and $\Delta, A \vdash B$ then $\Gamma, \Delta \vdash B$.
3. if $\Gamma \vdash A$, then there is a finite subset $\Delta \subset \Gamma$ such that $\Delta \vdash A$.

In the Hilbert system above, the so-called *deduction theorem* holds.

**Theorem 3.11** (Deduction theorem) *Let $\Gamma$ be a set of formulas and $A, B$ be formulas. Then, we have:*

$\Gamma, A \vdash B \Rightarrow \Gamma \vdash A \rightarrow B.$

The following theorem shows some theorems related to strong negation.

**Theorem 3.12** *Let A and B be any formula. Then,*

1. $\vdash A \vee \neg_* A$
2. $\vdash A \rightarrow (\neg_* A \rightarrow B)$
3. $\vdash (A \rightarrow B) \rightarrow ((A \rightarrow \neg_* B) \rightarrow \neg_* A)$

From Theorems 3.11 and 3.12 follows.

**Theorem 3.13** *For arbitrary formulas A and B, the following hold:*

1. $\vdash \neg_*(A \wedge \neg_* A)$
2. $\vdash A \leftrightarrow \neg_* \neg_* A$
3. $\vdash (A \wedge B) \leftrightarrow \neg_*(\neg_* A \vee \neg_* B)$
4. $\vdash (A \rightarrow B) \leftrightarrow (\neg_* A \vee B)$
5. $\vdash (A \vee B) \leftrightarrow \neg_*(\neg_* A \wedge \neg_* B)$.

Theorem 3.13 implies that by using strong negation and a logical connective other logical connectives can be defined as in classical logic. If $\tau = \{t, f\}$, with its operations appropriately defined, we can obtain classical propositional logic in which $\neg_*$ is classical negation.

Now, we provide some formal results of $P\tau$ including completeness and decidability.

**Lemma 3.9** *Let p be a propositional variable and $\mu$, $\lambda$, $\theta \in |\tau|$. Then, the following hold:*

1. $\vdash p_{\lambda \vee \mu} \rightarrow p_\lambda$
2. $\vdash p_{\lambda \vee \mu} \rightarrow p_\mu$
3. $\lambda \geq \mu$ *and* $\lambda \geq \theta \Rightarrow \vdash p_\lambda \rightarrow p_{\mu \vee \theta}$
4. $\vdash p_\mu \rightarrow p_{\mu \wedge \theta}$.
5. $\vdash p_\theta \rightarrow p_{\mu \wedge \theta}$.
6. $\lambda \leq \mu$ *and* $\lambda \leq \theta \Rightarrow \vdash p_{\mu \wedge \theta}$
7. $\vdash p_\mu \leftrightarrow p_{\mu \vee \mu}$, $\vdash p_\mu \leftrightarrow p_{\mu \wedge \mu}$
8. $\vdash p_{\mu \vee \lambda} \leftrightarrow p_{\lambda \vee \mu}$, $\vdash p_{\mu \wedge \lambda} \leftrightarrow p_{\lambda \wedge \mu}$
9. $\vdash p_{(\mu \vee \lambda) \vee \theta} \vee \rightarrow p_{\mu \vee (\lambda \vee \theta)}$, $\vdash p_{(\mu \wedge \lambda) \wedge \theta} \vee \rightarrow p_{\mu \wedge (\lambda \wedge \theta)}$
10. $\vdash p_{(\mu \vee \lambda) \wedge \mu} \rightarrow p_\mu$, $\vdash p_{(\mu \wedge \lambda) \vee \mu} \rightarrow p_\mu$
11. $\lambda \leq \mu \Rightarrow \vdash p_{\lambda \vee \mu} \rightarrow p_\mu$
12. $\lambda \vee \mu = \mu \Rightarrow \vdash p_\mu \rightarrow p_\lambda$
13. $\mu \geq \lambda \Rightarrow \forall \theta \in |\tau|$ $(\vdash p_{\mu \vee \theta} \rightarrow p_{\lambda \vee \theta}$ *and* $\vdash p_{\mu \wedge \theta} \rightarrow p_{\lambda \wedge \theta})$
14. $\mu \geq \lambda$ *and* $\theta \geq \varphi \Rightarrow \vdash p_{\mu \vee \theta} \rightarrow p_{\lambda \vee \varphi}$ *and* $p_{\mu \wedge \theta} \rightarrow p_{\lambda \wedge \varphi}$

15. $\vdash p_{\mu \wedge (\lambda \vee \theta)} \rightarrow p_{(\mu \wedge \lambda) \vee (\mu \wedge \theta)}$, $\vdash p_{\mu \vee (\lambda \wedge \theta)} \rightarrow p_{(\mu \vee \lambda) \wedge (\mu \vee \theta)}$
16. $\vdash p_\mu \wedge p_\lambda \leftrightarrow p_{\mu \wedge \lambda}$
17. $\vdash p_{\mu \vee \lambda} \rightarrow p_\mu \vee p_\lambda$.

Consider the complete lattice $\tau = N \cup \{\omega\}$, where $N$ is the set of natural numbers. The ordering on $\tau$ is the usual ordering on ordinals, restricted to the set $\tau$. Consider the set $\Gamma = \{p_0, p_1, p_2, ...\}$, where $p_\omega \notin \Gamma$. It is clear that $\Gamma \vdash p_\omega$, but an infinitary deduction is required to establish this.

**Definition 3.12** $\overline{\Delta} = \{A \in \mathbf{F} \mid \Delta \vdash A\}$

**Definition 3.13** $\Delta$ is said to be *trivial* iff $\overline{\Delta} = \mathbf{F}$ (i.e., every formula in our language is a syntactic consequence of $\Delta$); otherwise, $\Delta$ is said to be *non-trivial*. $\Delta$ is said to be *inconsistent* iff there is some formula $A$ such that $\Delta \vdash A$ and $\Delta \vdash \neg A$; otherwise, $\Delta$ is *consistent*.

From the definition of triviality, the next theorem follows:

**Theorem 3.14** *$\Delta$ is trivial iff $\Delta \vdash A \wedge \neg A$ (or $\Delta \vdash A$ and $\Delta \vdash \neg_* A$) for some formula $A$.*

**Theorem 3.15** *Let $\Gamma$ be a set of formulas, $A, B$ be any formulas, and $F$ be any complex formula. Then, the following hold.*

1. $\Gamma \vdash A$ *and* $\Gamma \vdash A \rightarrow B \Rightarrow \Gamma \vdash B$
2. $A \wedge B \vdash A$
3. $A \wedge B \vdash B$
4. $A, B \vdash A \wedge B$
5. $A \vdash A \vee B$
6. $B \vdash A \vee B$
7. $\Gamma, A \vdash C$ *and* $\Gamma, B \vdash C \Rightarrow \Gamma, A \vee B \vdash C$
8. $\vdash F \leftrightarrow \neg_* F$
9. $\Gamma, A \vdash B$ *and* $\Gamma, A \vdash \neg_* B \Rightarrow \Gamma \vdash \neg_* A$
10. $\Gamma, A \vdash B$ *and* $\Gamma, \neg_* A \vdash B \Rightarrow \Gamma \vdash B$.

Note here that the counterpart of Theorem 3.15 (10) obtained by replacing the occurrence of $\neg_*$ by $\neg$ is not valid.

Now, we are in a position to prove the soundness and completeness of $P\tau$. Our proof method for completeness is based on maximal non-trivial set of formulas; see Abe [62] and Abe, Akama and Nakamatsu [63]. da Costa, Subrahmanian and Vago [60] presented another proof using Zorn's Lemma.

**Theorem 3.16** (Soundness) *Let $\Gamma$ be a set of formulas and $A$ be any formula. $\mathscr{A}\tau$ is a sound axiomatization of $P\tau$, i.e., if $\Gamma \vdash A$ then $\Gamma \models A$.*

For proving the completeness theorem, we need some theorems.

**Theorem 3.17** *Let $\Gamma$ be a non-trivial set of formulas. Suppose that $\tau$ is finite. Then, $\Gamma$ can be extended to a maximal (with respect to inclusion of sets) non-trivial set with respect to* **F**.

**Theorem 3.18** *Let $\Gamma$ be a maximal non-trivial set of formulas. Then, we have the following:*

1. *if $A$ is an axiom of $P\tau$, then $A \in \Gamma$*
2. *$A, B \in \Gamma$ iff $A \wedge B \in \Gamma$*
3. *$A \vee B \in \Gamma$ iff $A \in \Gamma$ or $B \in \Gamma$*
4. *if $p_\lambda, p_\mu \in \Gamma$, then $p_\theta \in \Gamma$, where $\theta = max(\lambda, \mu)$*
5. *$\neg^k p_\mu \in \Gamma$ iff $\neg^{k-1} p_{\sim\mu} \in \Gamma$, where $k \geq 1$*
6. *if $A, A \rightarrow B \in \Gamma$, then $B \in \Gamma$*
7. *$A \rightarrow B \in \Gamma$ iff $A \notin \Gamma$ or $B \in \Gamma$.*

**Theorem 3.19** *Let $\Gamma$ be a maximal non-trivial set of formulas. Then, the characteristic function $\chi$ of $\Gamma$, that is, $\chi_\Gamma \rightarrow$ **2** is the valuation function of some interpretation $I : \mathbf{P} \rightarrow |\tau|$.*

Here is the completeness theorem for $P\tau$.

**Theorem 3.20** (Completeness) *Let $\Gamma$ be a set of formulas and $A$ be any formula. If $\tau$ is finite, then $\mathscr{A}\tau$ is a complete axiomatization for $P\tau$, i.e., if $\Gamma \models A$ then $\Gamma \vdash A$.*

The decidability theorem also holds for finite lattice.

**Theorem 3.21** (Decidability) *If $\tau$ is finite, then $P\tau$ is decidable.*

The completeness does not in general hold for infinite lattice. But, it holds for special case.

**Definition 3.14** (*Finite annotation property*) Suppose that $\Gamma$ be a set of formulas such that the set of annotated constants occurring in $\Gamma$ is included in a finite substructure of $\tau$ ($\Gamma$ itself may be infinite). In this case, $\Gamma$ is said to have the *finite annotation property*.

Note that if $\tau'$ is a substructure of $\tau$ then $\tau'$ is closed under the operations $\sim$, $\vee$ and $\wedge$. One can easily prove the following from Theorem 3.20.

**Theorem 3.22** (Finitary Completeness) *Suppose that $\Gamma$ has the finite annotation property. If $A$ is any formula such that $\Gamma \vdash A$, then there is a finite proof of $A$ from $\Gamma$.*

Theorem 3.20 tells us that even if the set of the underlying truth-values of $P\tau$ is infinite (countably or uncountably), as long as theories have the finite annotation property. The completeness result applied to them, i.e., $\mathscr{A}\tau$ is complete with respect to such theories.

In general, when we consider theories that do not possess the finite annotation property, it may be necessary to guarantee completeness by adding a new infinitary inference rule ($\omega$-rule), similar in spirit to the rule used by da Costa [64] in order to cope with certain models in a particular family of infinitary language. Observe that for such cases a desired axiomatization of $P\tau$ is not finitary.

From the classical result of compactness, we can state a version of the compactness theorem.

**Theorem 3.23** (Weak Compactness) *Suppose that $\Gamma$ has the finite annotation property. If $A$ is any formula such that $\Gamma \vdash A$, then there is a finite subset $\Gamma'$ of $\Gamma$ such that $\Gamma' \vdash A$.*

Annotated logics $P\tau$ provide a general framework, and can be used to reasoning about many different logics. Below we present some examples.

The set of truth-values $FOUR = \{t, f, \bot, \top\}$, with $\neg$ defined as: $\neg t = f$, $\neg f = t$, $\neg\bot = \bot$, $\neg\top = \top$. Four-valued logic based on *FOUR* was originally due to Belnap [7, 8] to model internal states in a computer.

Subrahmanian [58] formalized an annotated logic with *FOUR* as a foundation for paraconsistent logic programming; also see Blair and Subrahmanian [59]. Their annotated logic may be used for reasoning about inconsistent knowledge bases.

For example, we may allow logic programs to be finite collections of formulas of the form:

$$(A : \mu_0) \leftrightarrow (B_1 : \mu_1)\&...\&(B_n : \mu_n)$$

where $A$ and $B_i$ $(1 \leq i \leq n)$ are atoms and $\mu_j$ $(0 \leq j \leq n)$ are truth-values in *FOUR*.

Intuitively, such programs may contain "intuitive" inconsistencies–for example, the pair

$$((p : f), (p : t))$$

is inconsistent. If we append this program to a consistent program $P$, then the resulting union of these two programs may be inconsistent, even though the predicate symbols $p$ occurs nowhere in program $P$.

Such inconsistencies can easily occur in knowledge based systems, and should not be allowed to trivialize the meaning of a program. However, knowledge based systems based on classical logic cannot handle the situation since the program is trivial.

In Blair and Subrahmanian [59], it is shown how the four-valued annotated logic may be used to describe this situation. Later, Blair and Subrahmanian's annotated logic was extended as *generalized annotated logics* by Kifer and Subrahmanian [17].

There are also other examples which can be dealt with by annotated logics. The set of truth-values *FOUR* with negation defined as boolean complementation forms an annotated logic.

The unit interval [0, 1] of truth-values with $\neg x = 1 - x$ is considered as the base of annotated logic for qualitative or fuzzy reasoning. In this sense, probabilistic and fuzzy logics could be generalized as annotated logics.

The interval $[0, 1] \times [0, 1]$ of truth-values can be also used for annotated logics for evidential reasoning. Here, the assignment of the truth-value $(\mu_1, \mu_2)$ to proposition

$p$ may be thought of as saying that the degree of belief in $p$ is $\mu_1$, while the degree of disbelief is $\mu_2$. Negation can be defined as $\neg(\mu_1, \mu_2) = (\mu_2, \mu_1)$.

Note that the assignment of $[\mu_1, \mu_2]$ to a proposition $p$ by an interpretation $I$ does not necessarily satisfy the condition $\mu_1 + \mu_2 \leq 1$. This contrasts with probabilistic reasoning. Knowledge about a particular domain may be gathered from different experts (in that domain), and these experts may different views.

Some of these views may lead to a "strong" belief in a proposition; likewise, other experts may have a "strong" disbelief in the same proposition. In such a situation, it seems appropriate to report the existence of conflicting opinions, rather than use ad-hoc means to resolve this conflict.

As mentioned above, da Costa, Subrahmanian and Vago [60] investigated propositional annotated logics $P\tau$, and suggested their predicate extension $Q\tau$ (also denoted $Q\mathscr{T}$). We can look at the detailed formulation of $Q\tau$ in da Costa, Abe, and Subrahmanian [61]; also see Abe [62]. But, we here omit the exposition of $Q\tau$.

Our exposition of non-classical logics has finished. We only considered non-classical logics which are related to rough set theory. They serve as foundations for rough set logics as will be discussed Chap. 4.

## References


1. Kripke, S.: A complete theorem in modal logic. J. Symb. Log. **24**, 1–24 (1959)
2. Kripke, S.: Semantical considerations on modal logic. Acta Philos. Fenn. **16**, 83–94 (1963)
3. Kripke, S.: Semantical analysis of modal logic I. Z. für math. Logik und Grundl. der Math. **8**, 67–96 (1963)
4. Hughes, G., Cresswell, M.: A New Introduction to Modal Logic. Routledge, New York (1996)
5. Łukasiewicz, J.: On 3-valued logic 1920. In: McCall, S. (ed.) Polish Logic, pp. 16–18. Oxford University Press, Oxford (1967)
6. Kleene, S.: Introduction to Metamathematics. North-Holland, Amsterdam (1952)
7. Belnap, N.D.: A useful four-valued logic. In: Dunn, J.M., Epstein, G. (eds.) Modern Uses of Multi-Valued Logic, pp. 8–37. Reidel, Dordrecht (1977)
8. Belnap, N.D.: How a computer should think. In: Ryle, G. (ed.) Contemporary Aspects of Philosophy, pp. 30–55. Oriel Press (1977)
9. Anderson, A., Belnap, N.: Entailment: The Logic of Relevance and Necessity I. Princeton University Press, Princeton (1976)
10. Łukasiewicz, J.: Many-valued systems of propositional logic, 1930. In: McCall, S. (ed.) Polish Logic. Oxford University Press, Oxford (1967)
11. Zadeh, L.: Fuzzy sets. Inf. Control **8**, 338–353 (1965)
12. Fitting, M.: Bilattices and the semantics of logic programming. J. Log. Program. **11**, 91–116 (1991)
13. Fitting, M.: A theory of truth that prefers falsehood. J. Philos. Log. **26**, 477–500 (1997)
14. Ginsberg, M.: Multivalued logics. In: Proceedings of AAAI 1986, pp. 243–247. Morgan Kaufman, Los Altos (1986)
15. Ginsberg, M.: Multivalued logics: a uniform approach to reasoning in AI. Comput. Intell. **4**, 256–316 (1988)
16. Fitting, M.: Intuitionisic Logic, Model Theory and Forcing. North-Holland, Amsterdam (1969)
17. Kifer, M., Subrahmanian, V.S.: On the expressive power of annotated logic programs. In: Proceedings of the 1989 North American Conference on Logic Programming, pp. 1069–1089 (1989)

18. Kripke, S.: Outline of a theory of truth. J. Philos. **72**, 690–716 (1975)
19. Arieli, O., Avron, A.: Reasoning with logical bilattices. J. Log. Lang. Inf. **5**, 25–63 (1996)
20. Arieli, O., Avron, A.: The value of fur values. Artif. Intell. **102**, 97–141 (1998)
21. Heyting, A.: Intuitionism. North-Holland, Amsterdam (1952)
22. Kripke, S.: Semantical analysis of intuitionistic logic. In: Crossley, J., Dummett, M. (eds.) Formal Systems and Recursive Functions, pp. 92–130. North-Holland, Amsterdam (1965)
23. Akama, S.: The Gentzen-Kripke construction of the intermediate logic LQ. Notre Dame J. Form. Log. **33**, 148–153 (1992)
24. Dummett, M.: A propositional calculus with denumerable matrix. J. Symb. Log. **24**, 97–106 (1959)
25. Krisel, G., Putnam, H.: Eine unableitbarkeitsbeuwesmethode für den intuitinistischen Aussagenkalkul. Arch. für Math. Logik und Grundlagenforschung **3**, 74–78 (1967)
26. Akama, S., Murai, T., Kudo, Y.: Heyting-Brouwer rough set logic. In: Proceedings of KSE2013, Hanoi, pp. 135–145. Springer, Heidelberg (2013)
27. Akama, S., Murai, T., Kudo, Y.: Da Costa logics and vagueness. In: Proceedings of GrC2014, Noboribetsu, Japan (2014)
28. Nelson, D.: Constructible falsity. J. Symb. Log. **14**, 16–26 (1949)
29. Almukdad, A., Nelson, D.: Constructible falsity and inexact predicates. J. Symb. Log. **49**, 231–233 (1984)
30. Akama, S.: Resolution in constructivism. Log. et Anal. **120**, 385–399 (1987)
31. Akama, S.: Constructive predicate logic with strong negation and model theory. Notre Dame J. Form. Log. **29**, 18–27 (1988)
32. Akama, S.: On the proof method for constructive falsity. Z. für Math. Log. und Grundl. der Math. **34**, 385–392 (1988)
33. Akama, S.: Subformula semantics for strong negation systems. J. Philos. Log. **19**, 217–226 (1990)
34. Akama, S.: Constructive falsity: foundations and their applications to computer science. Ph.D. thesis, Keio University, Yokohama, Japan (1990)
35. Akama, S.: Nelson's paraconsistent logics. Log. Log. Philos. **7**, 101–115 (1999)
36. Wansing, H.: The Logic of Information Structures. Springer, Berlin (1993)
37. Nelson, D.: Negation and separation of concepts in constructive systems. In: Heyting, A. (ed.) Constructivity in Mathematics, pp. 208–225. North-Holland, Amsterdam (1959)
38. Rasiowa, H.: An Algebraic Approach to Non-Classical Logics. North-Holland, Amsterdam (1974)
39. Jaśkowski, S.: Propositional calculus for contradictory deductive systems (in Polish). Stud. Soc. Sci. Tor. Sect. A **1**, 55–77 (1948)
40. Jaśkowski, S.: On the discursive conjunction in the propositional calculus for inconsistent deductive systems (in Polish). Stud. Soc. Sci. Tor. Sect. A **8**, 171–172 (1949)
41. Kotas, J.: The axiomatization of S. Jaskowski's discursive logic. Stud. Log. **33**, 195–200 (1974)
42. da Costa, N.C.A.: On the theory of inconsistent formal systems. Notre Dame J. Form. Log. **15**, 497–510 (1974)
43. da Costa, N.C.A., Alves, E.H.: A semantical analysis of the calculi $C_n$. Notre Dame J. Form. Log. **18**, 621–630 (1977)
44. Anderson, A., Belnap, N., Dunn, J.: Entailment: The Logic of Relevance and Necessity II. Princeton University Press, Princeton (1992)
45. Routley, R., Plumwood, V., Meyer, R.K., Brady, R.: Relevant Logics and Their Rivals, vol. 1. Ridgeview, Atascadero (1982)
46. Dunn, J.M.: Relevance logic and entailment. In: Gabbay, D., Gunthner, F. (eds.) Handbook of Philosophical Logic, vol. III, pp. 117–224. Reidel, Dordrecht (1986)
47. Vasil'ev, N.A.: Imaginary Logic. Nauka, Moscow (1989). (in Russian)
48. Arruda, A.I.: A survey of paraconsistent logic. In: Arruda, A., da Costa, N., Chuaqui, R. (eds.) Mathematical Logic in Latin America, North-Holland, Amsterdam, pp. 1–41 (1980)
49. Asenjo, F.G.: A calculus of antinomies. Notre Dame J. Form. Log. **7**, 103–105 (1966)
50. Priest, G.: Logic of paradox. J. Philos. Log. **8**, 219–241 (1979)

51. Priest, G.: Paraconsistent logic. In: Gabbay, D., Guenthner, F. (eds.) Handbook of Philosophical Logic, 2nd edn, pp. 287–393. Kluwer, Dordrecht (2002)
52. Priest, G.: In Contradiction: A Study of the Transconsistent, 2nd edn. Oxford University Press, Oxford (2006)
53. Batens, D.: Inconsistency-adaptive logics and the foundation of non-monotonic logics. Log. et Anal. **145**, 57–94 (1994)
54. Batens, D.: A general characterization of adaptive logics. Log. et Anal. **173–175**, 45–68 (2001)
55. Batens, D.: Dynamic dialectical logics. In: Priest, G., Routley, R., Norman, J. (eds.) Paraconsistent Logic: Essay on the Inconsistent, pp. 187–217. Philosophia Verlag, München (1989)
56. Carnielli, W., Coniglio, M., Marcos, J.: Logics of formal inconsistency. In: Gabbay, D., Guenthner, F. (eds.) Handbook of Philosophical Logic, vol. 14, 2nd edn, pp. 1–93. Springer, Heidelberg (2007)
57. Carnielli, W., Marcos, J.: Tableau systems for logics of formal inconsistency. In: Abrabnia, H.R. (ed.) Proceedings of the 2001 International Conference on Artificial Intelligence, vol. II, pp. 848–852. CSREA Press (2001)
58. Subrahmanian, V.: On the semantics of quantitative logic programs. In: Proceedings of the 4th IEEE Symposium on Logic Programming, pp. 173–182 (1987)
59. Blair, H.A., Subrahmanian, V.S.: Paraconsistent logic programming. Theor. Comput. Sci. **68**, 135–154 (1989)
60. da Costa, N.C.A., Subrahmanian, V.S., Vago, C.: The paraconsistent logic $P\mathcal{T}$. Z. für Math. Log. und Grundl. der Math. **37**, 139–148 (1991)
61. da Costa, N.C.A., Abe, J.M., Subrahmanian, V.S.: Remarks on annotated logic. Z. für Math. Log. und Grundl. der Math. **37**, 561–570 (1991)
62. Abe, J.M.: On the foundations of annotated logics (in Portuguese). Ph.D. thesis, University of São Paulo, Brazil (1992)
63. Abe, J.M., Akama, S., Nakamatsu, K.: Introduction to Annotated Logics. Springer, Heidelberg (2015)
64. da Costa, N.C.A.: $\alpha$-models and the system $T$ and $T^*$. Notre Dame J. Form. Log. **14**, 443–454 (1974)

# Chapter 4
# Logical Characterizations of Rough Sets

**Abstract** This chapter introduces several logical characterizations of rough sets. We outline some approaches in the literature, including double Stone algebras, Nelson algebras and modal logics. We also discuss rough set logics, logics for reasoning about knowledge, and logics for knowledge representation.

## 4.1 Algebraic Approaches

One of the most basic approaches to rough sets is an *algebraic approach*, which provides mathematical characterization. There are several algebraic approaches, and the first one is due to Iwinski [1] in 1987.

A suitable algebra for rough sets is a *double Stone algebra* (DSA). Below we review several algebraic approaches. We here assume basics of algebras.

**Definition 4.1** *(Double Stone Algebra)* A double Stone algebra $\langle L, +, \cdot, *, ^+, 0, 1\rangle$ is an algebra of type (2, 2, 1, 1, 0, 0) such that:

(1) $\langle L, +, \cdot, 0, 1\rangle$ is a bounded distributive lattice,
(2) $x^*$ is the pseudocomplement of $x$, i.e., $y \leq x^* \Leftrightarrow y \cdot x = 0$,
(3) $x^+$ is the dual pseudocomplement of $x$, i.e., $y \geq x^+ \Leftrightarrow y + x = 1$.
(4) $x^* + x^{**} = 1,\ x^+ \cdot x^{++} = 0$.

Note that conditions (2) and (3) can be described as the following equations:

$$x \cdot (x \cdot y)^* = x \cdot y^*,$$
$$x + (x + y)^+ = x + y^+,$$
$$x \cdot 0^* = x,$$
$$x + 1^+ = x,$$
$$0^{**} = 0,$$
$$1^{++} = 1.$$

DSA is called *regular* if it additionally satisfies:

$$x \cdot x^+ \leq y + y^*$$

which is equivalent to $x^+ = y^+$ and $x^* = y^*$ imply $x = y$.



S. Akama et al., *Reasoning with Rough Sets*, Intelligent Systems Reference Library 142, https://doi.org/10.1007/978-3-319-72691-5_4

The *center* $B(L) = \{x^* \mid x \in L\}$ of $L$ is a subalgebra of $L$ and a Boolean algebra, in which $^*$ and $^+$ coincide with the Boolean complement $-$.

An element of the center of $L$ is called a *Boolean element*. The dense set $\{x \in L \mid x^* = 0\}$ of $L$ is denoted by $D(L)$ or simply $D$. For any $M \subseteq L$, $M^+$ is the set $\{x^+ \mid x \in M\}$.

Lemma 4.1 is a construction of regular double Stone algebras.

**Lemma 4.1** *Let* $\langle B, +, \cdot, -, 0, 1\rangle$ *be a Boolean algebra and* $F$ *be a not necessarily proper filter on* $B$. *Set*

$$\langle a, b\rangle^* = \langle -b, -b\rangle,$$
$$\langle a, b\rangle^+ = \langle -a, -a\rangle.$$

*Furthermore* $B(L) \simeq B$ *as Boolean algebras, and* $D(L) \simeq F$ *as lattices. Note that* $B(L) = \{\langle a, a\rangle \mid a \in B\}$, $D(L) = \{\langle a, 1\rangle \mid a \in F\}$.
*Conversely, if* $M$ *is a regular double Stone algebra,* $B = B(M)$, $F = D(M)^{++}$ *then the mapping which assigns to each* $x \in M$ *the pair* $\langle x^{++}, x^{**}\rangle$ *is an isomorphism* $M$ *and* $\langle B, F\rangle$.

It follows that each element $x$ of a regular double Stone algebra is uniquely described by the greatest Boolean element below $x$ and the smallest Boolean element above $x$.

Now, suppose $\langle U, \theta\rangle$ is an approximation space. Then, the classes of $\theta$ can be regarded as atoms of a complete subalgebra of the Boolean algebra $Sb(U)$. Conversely, any atomic complete subalgebra $B$ of $Sb(U)$ yields an equivalence relation $\theta$ on $U$, where this correspondence is a bijection. The elements of $B$ are $\emptyset$ and the unions of its associated equivalence relations.

If $a \in B$, then for every $X \subseteq U$, if $a \in X_u$, then $a \in X$, and the rough sets of the corresponding approximation space are the elements of the regular double Stone algebra $\langle B, F\rangle$, where $F$ is the filter of $B$ which is generated by the union of singleton element of $B$. If $\theta$ is the identity on $U$, then $F = \{U\}$.

Other algebraic approaches can be also found in the literature. For instance, Pomykala and Pomykala [2] showed that the collection $\mathscr{B}_\theta(U)$ of rough sets of $(U, \theta)$ can be made into a Stone algebra $(\mathscr{B}_\theta(U), +, \cdot, *, (\emptyset, \emptyset), (U, U))$ by defining,

$$(\underline{X}, \overline{X}) + (\underline{Y}, \overline{Y}) = (\underline{X} \cup \underline{Y}, \overline{X} \cup \overline{Y})$$
$$(\underline{X}, \overline{X}) \cdot (\underline{Y}, \overline{Y}) = (\underline{X} \cap \underline{Y}, \overline{X} \cap \overline{Y})$$
$$(\underline{X}, \overline{X})^* = (-\overline{X}, -\overline{X})$$

where for $X \subseteq U$, the complement of $Z$ in $U$ is denoted by $-Z$.

We can also characterize rough sets by Nelson algebras and three-valued Łukasiewicz algebras. We will discuss Nelson algebras for characterizing rough sets in Sect. 4.4.

## 4.2 Modal Logic and Rough Sets

It has been argued that modal logic offers foundations for rough set theory. In this line, Yao and Lin [3] provided a systematic study on the generalization of the modal logic approach. Consult Liau [4] for a comprehensive survey.

Let $apr = \langle U, R \rangle$ be an approximation space. Given an arbitrary set $A \subseteq U$, it may be impossible to describe $A$ precisely using the equivalence classes of $R$. In this case, one may characterize $A$ by a pair of lower and upper approximations:

$\underline{apr}(A) = \bigcup_{[x]_R \subseteq A} [x]_R,$
$\overline{apr}(A) = \bigcup_{[x]_R \cap A \neq \emptyset} [x]_R,$

where $[x]_R = \{y \mid xRy\}$ is the equivalence class containing $x$. The pair $\langle \underline{apr}(A), \overline{apr}(A) \rangle$ is the rough set with repect to $A$.

Since the lower approximation $\underline{apr}(A)$ is the union of all the elementary sets which are subsets of $A$ and the $\overline{apr}(A)$ is the union of all the elementary sets which have a non-empty intersection with $A$, they can be described as follows:

$$\begin{aligned}
\underline{apr}(A) &= \{x \mid [x]_R \subseteq A\} \\
&= \{x \in U \mid for\ all\ y \in U, xRy implies\ y \in A\} \\
\overline{apr}(A) &= \{x \mid [x]_R \cap A = \emptyset\} \\
&= \{x \in U \mid there\ exists\ a\ y \in U\ such\ that\ xRy\ and\ y \in A\}
\end{aligned}$$

These interpretations are closely related to those of the necessity and possibility operators in modal logic.

Here, we list some properties of $\underline{apr}$ and $\overline{apr}$. For any subsets $A, B \subseteq U$, we have the following about $\underline{apr}$:

(AL1) $\underline{apr}(A) = \sim \overline{apr}(\sim A)$
(AL2) $\underline{apr}(U) = U$
(AL3) $\underline{apr}(A \cap B) = \underline{apr}(A) \cap \underline{apr}(B)$
(AL4) $\underline{apr}(A \cup B) \supseteq \underline{apr}(A) \cup \underline{apr}(B)$
(AL5) $A \subseteq B \Rightarrow \underline{apr}(A) \subseteq \underline{apr}(B)$
(AL6) $\underline{apr}(\emptyset) = \emptyset$
(AL7) $\underline{apr}(A) \subseteq A$
(AL8) $A \subseteq \underline{apr}(\overline{apr}(A))$
(AL9) $\underline{apr}(A) \subseteq \underline{apr}(\underline{apr}(A))$
(AL10) $\overline{apr}(A) \subseteq \underline{apr}(\overline{apr}(A))$

and the following about $\overline{apr}$:

(AU1) $\overline{apr}(A) = \sim \underline{apr}(\sim A)$
(AU2) $\overline{apr}(\emptyset) = \emptyset$
(AU3) $\overline{apr}(A \cup B) = \overline{apr}(A) \cup \overline{apr}(B)$
(AU4) $\overline{apr}(A \cap B) \subseteq \overline{apr}(A) \cap \overline{apr}(B)$
(AU5) $A \subseteq B \Rightarrow \overline{apr}(A) \subseteq \overline{apr}(B)$
(AU6) $\overline{apr}(U) = U$

(AU7) $A \subseteq \overline{apr}(A)$
(AU8) $\overline{apr}(\underline{apr}(A)) \subseteq A$
(AU9) $\overline{apr}(\overline{apr}(A)) \subseteq \overline{apr}(A)$
(AU10) $\overline{apr}(Apr(A)) \subseteq \underline{apr}(A)$

where $\sim A = U - A$ denotes the set complement of $A$. Moreover, lower and upper approximations obey properties:

(K) $\underline{apr}(\sim A \cup B) \subseteq \sim \underline{apr}(A) \cup \underline{apr}(B)$
(ALU) $\underline{apr}(A) \subseteq \overline{apr}(A)$

Properties (AL1) and (AU1) state that two approximation operators are dual operators. In fact, properties with the same number may be regarded as dual properties.

However, these properties are not independent. For example, properties (AL3) implies properties (AL4). Properties (AL9), (AL10), (AU9) and (AU10) are expressed in terms of set inclusion.

Rough sets described above are called *Pawlak rough sets*. They are constructed from an equivalence relation. The pair of lower and upper approximations may be interpreted as two operators $\underline{apr}$ and $\overline{apr}$ of $U$.

Pawlak rough set model can be interpreted by the notions of topological space and topological Boolean algebra, where the operators $\underline{apr}$ and $\overline{apr}$ can be used together with the usual set-theoretic operators $\sim$, $\cap$ and $\cup$.

Since the modal logic S5 can be understood algebraically by topological Boolean algebra, it is natural to expect that there is a connection of Pawlak rough set and modal logic S5.

Now, we describe the presentation of modal logic as in Chap. 3. Let $\Phi$ be a non-empty set of propositions, which is generated by a finite set of logical symbols $\wedge, \vee, \neg, \rightarrow$, modal operators $\Box$, $\Diamond$, propositional constants $\top, \bot$, and infinitely enumerable set $P = \{\phi, \psi, ...\}$ of propositional variables. $\Box\phi$ states that $\phi$ is necessary and $\Diamond\phi$ states that $\phi$ is possible.

Let $W$ be a non-empty set of possible worlds and $R$ be a binary relation called accessibility relation on $W$. The pair $(W, R)$ is called a *Kripke frame*. An interpretation in $(W, R)$ is a function $v : W \times P \rightarrow \{true, false\}$, which assigns a truth value for each propositional variable with respect to each particular world $w$. By $v(w, a) = true$, we mean that the proposition $a$ is true in the interpretation $v$ in the world, written $w \models_v a$.

We extend $v$ for all propositions in the usual way. We define $v^* : W \times \Phi \rightarrow \{true, false\}$ as follows:

(m0) for $a \in P$, $w \models_{v^*} a$ iff $w \models_v a$
(m1) $w \not\models_{v^*} \bot$, $w \models_{v^*} \top$
(m2) $w \models_{v^*} \phi \wedge \psi$ iff $w \models_{v^*} \phi$ and $w \models_{v^*} \psi$
(m3) $w \models_{v^*} \phi \vee \psi$ iff $w \models_{v^*} \phi$ or $w \models_{v^*} \psi$
(m4) $w \models_{v^*} \phi \rightarrow \psi$ iff $w \not\models_{v^*} \phi$ or $w \models_{v^*} \psi$
(m5) $w \models_{v^*} \neg\phi$ iff $w \not\models_{v^*} \phi$
(m6) $w \models_{v^*} \Box\phi$ iff $\forall w' \in W(wRw' \Rightarrow w' \models_{v^*} \phi)$
(m7) $w \models_{v^*} \Diamond\phi$ iff $\exists w' \in W(wRw'$ and $w' \models_{v^*} \phi)$.

Here, $\not\models$ means that not $\models$. For simplicity, we will write $w \models_v \phi$, and we will drop $v$ when it is clear from context. The necessity and possibility are dual in the sense that the following hold:

$\Box\phi =_{\text{def}} \neg\Diamond\neg\phi,$
$\Diamond\phi =_{\text{def}} \neg\Box\neg\phi.$

If the accessibility relation $R$ is an equivalence relation, then the modal logic is S5.

We can characterize a proposition by the set of possible worlds in which it is true using a valuation function $v$. Based on the idea, we can define a mapping $t : \Phi \rightarrow 2^W$ as follows:

$$t(\phi) = \{w \in W : w \models \phi\}$$

Here, we call $t(\phi)$ the *truth set* of the proposition $\phi$. In the truth set representation, we have:

(s1) $t(\bot) = \emptyset,\ t(\top) = W$
(s2) $t(\phi \wedge \psi) = t(\phi) \cap t(\psi)$
(s3) $t(\phi \vee \psi) = t(\phi) \cap t(\psi)$
(s4) $t(\phi \rightarrow \psi) = \sim t(\phi) \cup t(\psi)$
(s5) $t(\neg\phi) = \sim t(\phi)$
(s6) $t(\Box\phi) = \underline{apr}(t(\phi))$
(s7) $t(\Diamond\phi) = \overline{apr}(t(\phi))$

The last two properties can be derived as follows:

$$\begin{aligned} t(\Box\phi) &= \{w \in W : w \models \Box\phi\} \\ &= \{w \in W : \forall w'(wRw' \Rightarrow w' \models \phi)\} \\ &= \{w \in W : \forall w'(wRw' \Rightarrow w' \in t(\phi))\} \\ &= \underline{apr}(t(\phi)) \\ t(\Diamond\phi) &= \{w \in W : w \models \Diamond\phi\} \\ &= \{w \in W : \exists w'(wRw' \text{ and } w' \models \phi)\} \\ &= \{w \in W : \exists w'(wRw' \text{ and } w' \in t(\phi))\} \\ &= \overline{apr}(t(\phi)) \end{aligned}$$

In the truth set interpretation above, approximation operators in Pawlak rough set model are related to modal operators in S5. There are many modal systems according to the properties on the accessibility relation in Kripke models.

Following Chellas's naming, axioms of modal logic corresponding to (AK), (ALU), (AL7)–(AL10) are given by:

(K) $\Box(\phi \rightarrow \psi) \rightarrow (\Box\phi \rightarrow \Box\psi)$
(D) $\Box\phi \rightarrow \Diamond\phi$
(T) $\Box\phi \rightarrow \phi$
(B) $\phi \rightarrow \Box\Diamond\phi$
(4) $\Box\phi \rightarrow \Box\Box\phi$
(5) $\Diamond\phi \rightarrow \Box\Diamond\phi.$

Pawlak rough set model can be characterized by modal logic S5. In fact, the standard logical operators are interpreted by usual set-theoretic operators and modal operators by rough set operators.

Since different systems of modal logic can be obtained by using various types of accessibility relations. It is thus possible to construct different rough set models by means of Kripke models for modal logic.

Yao and Lin worked out the idea by generalizing Pawlak's rough set models to provide different rough set models depending on the accessibility relation in Kripke models in Yao and Lin [3]. Their work seems to be of special interest of foundations for rough set theory. We simply review their work here.

Given a binary relation $R$ and two elements $x, y \in U$, if $xRy$ then we say that $y$ is $R$-related to $x$. A binary relation can be represented by a mapping $r : U \to 2^U$:

$r(x) = \{y \in U \mid xRy\}$

Here, $r(x)$ consists of all $R$-related elements of $x$. We then define two unary set-theoretic operators $\underline{apr}$ and $\overline{apr}$:

$$\begin{aligned}
\underline{apr}(A) &= \{x : r(x) \subseteq A\} \\
&= \{x \in U \mid \forall y \in U(xRy \Rightarrow y \in A)\} \\
\overline{apr}(A) &= \{x : r(x) \cap A \neq \emptyset\} \\
&= \{x \in U \mid \exists y \in U(xRy \text{ and } y \in A)\}
\end{aligned}$$

The set $\underline{apr}(A)$ consists of those elements where $R$-related elements are all in $A$, and $\overline{apr}(A)$ consists of those elements such that at least one of whose $R$-related elements is in $A$.

The pair $(\underline{apr}(A), \overline{apr}(A))$ denotes the *generalized rough set* of $A$ induced by $R$. Operators $\underline{apr}, \overline{apr} : 2^U \to 2^U$ are referred to as the generalized rough set operators. The induced system $(2^U, \cap, \cup, \sim, \underline{apr}, \overline{apr})$ is called an *algebraic rough set model*.

The equations (s6) and (s7) hold for generalized rough set operators. When $R$ is an equivalence relation, generalized rough set operators reduce to the operators in Pawlak rough set model.

We are now ready to present the classification of algebraic rough set models. First of all, we do not expect that generalized rough set operators satisfy all the properties in Pawlak rough set models. But properties (AL1)–(AL5) and (AU1)–(AU5) hold in any rough set model.

In modal logic, properties corresponding to (AK) and (AL6)–(AL10) yield different systems, and we can use these properties for various rough set models. Here, we relabel some properties used in modal logic:

(K) $\underline{apr}(\sim A \cup B) \subseteq \sim \underline{apr}(A) \cup \underline{apr}(B)$
(D) $\underline{apr}(A) \subseteq \overline{apr}(A)$
(T) $\underline{apr}(A) \subseteq A$
(B) $A \subseteq \underline{apr}(\overline{apr}(A))$
(4) $\underline{apr}(A) \subseteq \underline{apr}(\underline{apr}(A))$
(5) $\overline{apr}(A) \subseteq \underline{apr}(\overline{apr}(A))$

We need certain conditions on the binary relation $R$ to construct a rough set model so that the above properties hold.

A relation $R$ is serial if for all $x \in U$ there exists a $y \in U$ such that $xRy$. A relation is reflexive if for all $x \in U, xRx$ holds. A relation is symmetric if for all $x, y \in U, xRy$ implies $yRx$. A relation is transitive if for all $x, y, z \in U, xRy$ and $yRz$ imply $xRz$. A relation is Euclidian if for all $x, y, z \in U, xRy$ and $xRz$ imply $yRz$. The corresponding approximation operators are as follows:

| | |
|---|---|
| (serial) | for all x $\in U, r(x) \neq \emptyset$ |
| (reflexive) | for all x $\in U, x \in r(x)$ |
| (symmetric) | for all x, $y \in U$, if $x \in r(y)$, then y $\in r(x)$ |
| (transitive) | for all x,y $\in U$, if $y \in r(x)$, then $r(y) \subseteq r(x)$ |
| (Euclidian) | for all x,y $\in U$, if $y \in r(x)$, then $r(x) \subseteq r(y)$ |

Theorem 4.1 says rough set models with specific properties:

**Theorem 4.1** *The following relationships hold:*

1. *A serial rough set model satisfies (D).*
2. *A reflexive rough set model satisfies (T).*
3. *A symmetric rough set model satisfies (B).*
4. *A transitive rough set model satisfies (4).*
5. *A Euclidean rough set model satisfies (5).*

*Proof* (1) In a serial rough set model, for any $x \in \underline{apr}(A)$, we have $r(x) \subseteq A$ and $r(x) \neq \emptyset$. This implies $r(x) \cap A \neq \emptyset$., i.e., $x \in \overline{apr}(A)$. Thus, property (D) holds in a serial rough set model.
(2) In a reflexive rough set model, for any $x \in U, xRx$ implies $x \in r(x)$. Suppose $x \in \underline{apr}(A)$, which is equivalent to $r(x) \subseteq A$. Combining $x \in r(x)$ and $r(x) \subseteq A$, we have $x \in A$. Thus, property (T) holds.
(3) In a symmetric rough set model, suppose $x \in A$. By the symmetry of $R$, for all $y \in r(x)$, we have $x \in r(y)$, i.e., $x \in r(y) \cap A$. This implies that for all $y \in r(x), y \in \overline{apr}(A)$. Hence, $r(x) \subseteq \overline{apr}(A)$. It means that $x \in \underline{apr}(\overline{apr}(A))$. Therefore, property (B) holds.
(4) In a transitive rough set model, suppose $x \in \underline{apr}(A)$, i.e., $r(x) \subseteq A$. Then, for all $y \in r(x), r(y) \subseteq r(x) \subseteq A$, which is equivalent to say that for all $y \in r(x), y \in \underline{apr}(A)$. Thus, $r(x) \subseteq \underline{apr}(A)$ and in turn $x \in \underline{apr}(\underline{apr}(A))$. That is, property (4) holds.
(5) Conider an Euclidean rough set model. Suppose $x \in \overline{apr}(A)$, i.e., $r(x) \cap A \neq \emptyset$. By the Euclidean property of $R$, for all $y \in r(x), r(x) \subseteq r(y)$. Combining this with the assumption $r(x) \cap A \neq \emptyset$, we can conclude that for all $y \in r(x), y \in \overline{apr}(A)$. This is equivalent to say $r(x) \subseteq \overline{apr}(A)$. Therefore, (5) holds.

The five properties of a binary relation, namely the serial, reflexive, symmetric, transitive and Euclidean properties, induce five properties of the approximation operators, namely

serial: property (D) holds.
reflexive: property (T) holds.
symmetric: property (B) holds.
transitive: property (4) holds.
Euclidean: property (5) holds.

Combining these properties, it is possible to construct more rough set models. As a consequence of Theorem 4.1, we have:

**Theorem 4.2** *The following hold:*

1. *A T = KT rough set model satisfies (K), (T).*
2. *A B = KB rough set model satisfies (K), (B).*
3. *A S4 = KT4 rough set model satisfies (K), (T), and (4).*
4. *A S5 = KT5 (Pawlak) rough set model satisfies (K), (T), and (5).*

We can construct a rough set model corresponding to a normal modal system. Note that S5 (Pawlak) model is the strongest model. Yao and Lin's work established the connection of (normal) modal logic and rough set theory.

## 4.3 Graded and Probabilistic Modal Logics and Rough Sets

Yao and Lin generalized their modal logic approaches for graded and probabilistic modal logic. Here we review their approach.

Yao and Lin's work has been expanded in various ways. They presented graded and probabilistic versions. Because the quantitative information about the degree of overlap of $r(x)$ and $A$ is not considered in modal logic approach, such extensions are interesting.

*Graded modal logics* extend modal logic by introducing a family of graded modal operators $\Box_n$ and $\Diamond_n$, where $n \in N$ and $N$ is the set of natural numbers; see Fattorosi-Barnaba et al. [5–7]. These operators can be interpreted as follows:

(gm6) $w \models \Box_n \phi$ iff $|r(w)| - |t(\phi) \cap r(w)| \leq n$
(gm7) $w \models \Diamond_n \phi$ iff $|t(\phi) \cap r(w)| > n$.

where $|\cdot|$ denotes the cardinality of a set. $t(\phi)$ is the set of possible worlds in which $\phi$ is true and $r(w)$ is the set of possible worlds accessible from $w$.

The interpretations of $\Box_n \phi$ is that $\phi$ is false in at most $n$ possible worlds accesible from $w$, and the interpretaion of $\Diamond_n \phi$ is that $\phi$ is true in more than $n$ possible worlds accessible from $n$.

Graded necessity and possibility operators are dual:

$$\Box_n \phi =_{\text{def}} \neg \Diamond_n \neg \phi$$
$$\Diamond_n \phi =_{\text{def}} \neg \Box_n \neg \phi.$$

If $n = 0$, they reduce to normal modal operators.

$\Box\phi =_{\text{def}} \neg\Diamond\neg\phi$
$\Diamond\phi =_{\text{def}} \neg\Box\neg\phi.$

Further, we introduce a new graded modal operator $\Diamond!_n$, defined as $\Diamond!_n\phi = \Box_0\neg\phi$ and $\Diamond!_n\phi = \Diamond_{n-1}\phi \wedge \neg\Diamond_n\phi$ for $n > 0$. The interpretation of $\Diamond!_n\phi$ is that $\phi$ is true in exactly $n$ possible worlds accessible from $w$.

We can formalize graded modal systems. The basic graded modal logic is called Gr(K), in which axiom (K) is replaced by the following three axioms:

(GK1) $\Box_0(\phi \rightarrow \psi) \rightarrow (\Box_n\phi \rightarrow \Box_n\psi)$
(GK2) $\Box_n\phi \rightarrow \Box_{n+1}\phi$
(GK3) $\Box_0\neg(\phi \wedge \psi) \rightarrow ((\Diamond!_n\phi \wedge \Diamond!_m\psi) \rightarrow \Diamond_{n+m}(\phi \vee \psi))$

The graded modal logic Gr(T) is obtained by adding (GT) to Gr(K), and Gr(S5) is obtained by adding (G5) to Gr(T). The two axioms (GT) and (G5) are defined as follows:

(GT) $\Box_0\phi \rightarrow \phi$
(G5) $\Diamond_n\phi \rightarrow \Box_0\Diamond_n\phi.$

Based on graded modal logic, we introduce the notion of graded rough sets. Given the universe $U$ and a binary relation $R$ on $U$, a family of graded rough set operators are defined as:

$\underline{apr}_n(A) = \{x \mid \mid r(x) \mid - \mid A \cap r(x) \mid \leq n\}$
$\overline{apr}_n(A) = \{x \mid \mid A \cap r(x) \mid > n\}.$

An element of $U$ belongs to $\underline{apr}_n(A)$ if at most $n$ of its $R$-related elements are not in $A$, and belongs to $\overline{apr}_n(A)$ if more than $n$ of its $R$-related elements are in $A$. We establish a link between graded modal logics and rough sets:

(gs6) $t(\Box_n\phi) = \underline{apr}_n(t(\phi))$
(gs7) $t(\Diamond_n\phi) = \overline{apr}_n(t(\phi))$

We can interpret graded rough set operators in terms of graded modal operators. Independent of the types of binary relations, graded rough set operators satisfy the following properties:

(GL0) $\underline{apr}(A) = \underline{apr}_0(A)$
(GL1) $\underline{apr}_n(A) = \sim \overline{apr}_n(\sim A)$
(GL2) $\underline{apr}_n(A) = U$
(GL3) $\underline{apr}_n(A \cap B) \subseteq \underline{apr}_n(A) \cap \underline{apr}_n(A)$
(GL4) $\underline{apr}_n(A \cup B) \supseteq \underline{apr}_n(A) \cup \underline{apr}_n(A)$
(GL5) $A \subseteq B \Rightarrow \underline{apr}_n(A) \subseteq \underline{apr}_n(B)$
(GL6) $n \geq m \Rightarrow \underline{apr}_n(A) \supseteq \underline{apr}_m(A)$
(GU0) $\overline{apr}(A) = \overline{apr}_0(A)$
(GU1) $\overline{apr}_n(A) = \sim \underline{apr}_n(\sim A)$
(GU2) $\overline{apr}_n(\emptyset) = \emptyset$
(GU3) $\overline{apr}_n(A \cup B) \supseteq \overline{apr}_n(A) \cup \overline{apr}_n(B)$
(GU4) $\overline{apr}_n(A \cap B) \subseteq \overline{apr}_n(A) \cap \overline{apr}_n(B)$

(GU5) $A \subseteq B \Rightarrow \overline{apr}_n(A) \subseteq \overline{apr}_n(B)$
(GU6) $n \geq m \Rightarrow \overline{apr}_n(A) \subseteq \overline{apr}_m(A)$

Properties (GL0) and (GU0) show the relationship between graded rough set operators and normal rough set operators.

Properties (GL1)–(GL5) and (GU1)–(GU5) correspond to properties (AL1)–(AL5) and (AU1)–(AU5) of algebraic rough sets.

For properties (GL3) and (GU3), set equality is replaced by set inclusion.

Properties (GL6) and (GU6) characterize the relationships between graded modal operators. In fact, (GL6) corresponds to a graded version of (GK2) of graded modal logic. Properties corresponding to (GK1) and (GK3) can be easily constructed.

It is possible to construct different graded rough set models based on the properties satisfied by the binary relation. If the binary relation $R$ is an equivalence relation, we obtain the graded version of Pawlak rough sets.

The operators in graded Pawlak rough sets satisfy properties corresponding to axioms of graded modal logic:

(GD) $\underline{apr}_0(A) \subseteq \overline{apr}_0(A)$
(GT) $\underline{apr}_0(A) \subseteq A$
(GB) $A \subseteq \underline{apr}_0(\overline{apr}_0(A))$
(G4) $\underline{apr}_n(A) \subseteq \underline{apr}_0(\underline{apr}_n(A))$
(G5) $\overline{apr}_n(A) \subseteq \underline{apr}_0(\overline{apr}_n(A))$

We turn to probabilistic rough sets. Although we only use the *absolute* number of possible worlds accessible from a world $w$ and in which a proposition $\phi$ is true (false) in the definition of graded modal operators, the size of $r(w)$ is not considered. In probabilistic modal logic, all such information will be used.

Let $(W, R)$ be a frame. For each $w \in W$, we define a probabilistic function $P_w : \Phi \to [0, 1]$:

$$P_w(\phi) = \frac{|t(\phi) \cap r(w)|}{|r(w)|}$$

where $t(\phi)$ is the set of possible worlds in which $\phi$ is true, and $r(w)$ is the set of possible worlds accessible from $w$. Here, we implicitly assume that $R$ is at least serial, i.e., for all $w \in W$, $| r(w) | \geq 1$. Then, we define a family of probabilistic modal logic operators for $\alpha \in [0, 1]$:

$$\text{(pm6) } w \models \Box_\alpha \phi \text{ iff } P_w(\phi) \geq 1 - \alpha$$
$$\text{iff } \frac{|t(w) \cap r(w)|}{|r(w)|} \geq 1 - \alpha$$
$$\text{(pm7) } w \models \Diamond_\alpha \phi \text{ iff } P_w(\phi) > \alpha$$
$$\text{iff } \frac{|t(w) \cap r(w)|}{|r(w)|} > \alpha$$

These probabilistic modal operators are dual, i.e.,

$\Box_\alpha \phi =_{\text{def}} \neg \Diamond_\alpha \neg \phi$
$\Diamond_\alpha \phi =_{\text{def}} \neg \Box_\alpha \neg \phi.$

If $\alpha = 0$, then they agree to normal modal operators:

$\Box_0\phi =_{\text{def}} \Box\phi$
$\Diamond_0\phi =_{\text{def}} \Diamond\phi$.

The definition of probabilistic modal operators is consistent with that of Murai et al. [8, 9]. It is also a special case of the probabilistic Kripke model of Fattorosi-Barnaba and Amati in [10].

In fact, the probabilistic modal operators are related to the graded modal operators. If both sides of inequalities in (gm6) and (gm7) are divided by $|r(w)|$, and $n/\ |r(w)|$ is replaced by $\alpha$, the probabilistic modal operators are obtained.

But, these operators are different. Consider two possible worlds $w, w' \in W$ with $|r(w) \cap t(\phi)| = |r(w') \cap t(\phi)| = 1$ and $|r(w)| \neq |r(w')|$. We have:

$w \models \Diamond_0\phi$
$w' \models \Diamond_0\phi$

and

$w \models \neg\Diamond_n\phi$
$w' \models \neg\Diamond_n\phi$

for $n \geq 1$. That is, evaluations of $\Diamond_n\phi$ are the same in both worlds $w$ and $w'$. The difference in the size of $r(w)$ and $r(w')$ is reflected by operators $\Box_n$. On the other hand, since $1/\ |r(x)| \neq 1/\ |r(y)|$, evaluations of both $\Diamond_\alpha\phi$ and $\Box_\alpha$ will be different in worlds $w$ and $w'$.

We can then define probabilistic rough sets. Let $U$ be the universe and $R$ be a binary relation on $U$. We define a family of probabilistic rough set operators.

$$\underline{apr}_\alpha(A) = \{x \mid \frac{|A \cap r(x)|}{|r(x)|} \geq 1 - \alpha\}$$
$$\overline{apr}_\alpha(A) = \{x \mid \frac{|A \cap r(x)|}{|r(x)|} > \alpha\}.$$

With this definition, we can establish the connections between probabilistic modal logic and probabilistic rough sets:

(ps6) $t(\Box_\alpha\phi) = \underline{apr}_\alpha(t(\phi))$
(ps7) $t(\Diamond_\alpha\phi) = \overline{apr}_\alpha(t(\phi))$.

By definition, for a serial binary relation and $\alpha \in [0, 1]$, probabilistic rough set operators satisfy the following properties:

(PL0) $\underline{apr}(A) = \underline{apr}_0(A)$
(PL1) $\underline{apr}_\alpha(A) = \sim \overline{apr}_\alpha(\sim A)$
(PL2) $\underline{apr}_\alpha(U) = U$
(PL3) $\underline{apr}_\alpha(A \cap B) \subseteq \underline{apr}_\alpha \cap \underline{apr}_\alpha(B)$
(PL4) $\underline{apr}_\alpha(A \cup B) \subseteq \underline{apr}_\alpha \cup \underline{apr}_\alpha(B)$
(PL5) $A \subseteq B \Rightarrow \underline{apr}_\alpha(A) \subseteq \underline{apr}_\alpha(B)$
(PL6) $\alpha \geq \beta \Rightarrow \underline{apr}_\alpha(A) \supseteq \underline{apr}_\beta(A)$

(PU0) $\overline{apr}(A) = \overline{apr}_0(A)$
(PU1) $\overline{apr}_\alpha(A) = \sim \underline{apr}_\alpha(\sim A)$
(PU2) $\overline{apr}_\alpha(\emptyset) = \emptyset$
(PU3) $\overline{apr}_\alpha(A \cup B) \supseteq \overline{apr}_\alpha(A) \cup \overline{apr}_\alpha(B)$
(PU4) $\overline{apr}_\alpha(A \cap B) \subseteq \overline{apr}_\alpha(A) \cap \overline{apr}_\alpha(B)$
(PU5) $A \subseteq B \Rightarrow \overline{apr}_\alpha(A) \subseteq \overline{apr}_\alpha(B)$
(PU6) $\alpha \geq \beta \Rightarrow \overline{apr}_\alpha(A) \subseteq \overline{apr}_\beta(A)$

They are counterparts of the properties of graded rough set operators. Moreover, for $0 \leq \alpha < 0.5$,

(PD) $\underline{apr}_\alpha(A) \subseteq \overline{apr}_\alpha(A)$

which may be interpreted as a probabilistic version of axiom (D).

Probabilistic rough sets were first introduced by Wong and Ziarko [11] using probabilistic functions. Murai et al. also proposed similar probabilistic modal operators in [8, 9], which will be presented in Chap. 5.

## 4.4 Nelson Algebras and Rough Sets

Although rough set can be interpreted in terms of double Stone algebras, there are other algebraic approaches. One of the notable approaches is Pagliani's algebraic characterization based on *Nelson algebras* in Pagliani [12, 13].

Nelson algebras give an algebraic semantics for constructive logic with strong negation; see Rasiowa [14] and Vakarelov [15]. In 1996, Pagliani studied relations of Nelson algebras and rough sets in [12].

Recently, in 2013, Järvinen et al. proved an algebraic completeness for constructive logic with strong negation by using finite rough set-based Nelson algebras determined by quasiorders in [13].

We here present Pagliani's approach following [13]. Before doing it, we briefly state preliminary concepts.

A *Kleene algebra* is a structure $(A, \vee, \wedge, \sim, 0, 1)$ such that $A$ is a 0, 1-bounded distributed lattice and for all $a, b \in A$:

(K1) $\sim\sim a = a$
(K2) $a \leq b$ iff $\sim b \leq \sim a$
(K3) $a \wedge \sim a \leq b \vee \sim b$.

A *Nelson algebra* $(A, \vee, \wedge, \rightarrow, \sim, 0, 1)$ is a Kleene algebra $(A, \vee, \wedge, \sim, 0, 1)$ satisfying the following:

(N1) $a \wedge c \leq \sim a \vee b$ iff $c \leq a \rightarrow b$
(N2) $(a \wedge b) \rightarrow c = a \rightarrow (b \rightarrow c)$.

In Nelson algebras, an operation $\neg$ can be defined as $\neg a = a \rightarrow 0$. The operation $\rightarrow$ is called the *weak relative pseudocomplementation*, $\sim$ the strong negation, and $\neg$ the weak negation, respectively.

A Nelson algebra is *semi-simple* if $a \vee \neg a = 1$ for all $a \in A$. It is well known that semi-simple Nelson algebras coincide with three-valued Łukasiewicz algebras (cf. Iturrioz [16]) and regular double Stone algebras.

An element $a^*$ in a lattice $L$ with 0 is called a *pseudocomplement* of $a \in L$, if $a \wedge x = 0 \Leftrightarrow x \leq a^*$ for all $x \in L$. If a pseudocomplement of $a$ exists, then it is unique.

A *Heyting algebra* $H$ is a lattice with 0 such that for all $a, b \in H$, there is a greatest element $x$ of $H$ with $a \wedge x \leq b$.This element is called the *relative pseudo-complement* of $a$ with respect to $b$, denoted $a \Rightarrow b$. A Heyting algebra is represented as $(H, \vee, \wedge, \Rightarrow, 0, 1)$, where the pseudocomplement of $a$ is $a \Rightarrow 0$.

Let $\Theta$ be a Boolean congruence on a Heyting algebra $H$. Sendlewski [17] showed the construction of Nelson algebras from the pairs of Heyting algebras. In fact, the set of pairs

$$N_\Theta(H) = \{(a, b) \in H \times H \mid a \wedge b = 0 \textit{ and } a \vee b \Theta 1\}$$

can be seen as a Nelson algebra, if we add the operations:

$$(a, b) \vee (c, d) = (a \vee c, b \wedge d)$$
$$(a, b) \wedge (c, d) = (a \wedge c, b \vee d)$$
$$(a, b) \rightarrow (c, d) = (a \Rightarrow c, a \wedge d)$$
$$\sim (a, b) = (b, a)$$

Note that (0,1) is the 0-element and (1,0) is the 1-element. In the right-hand side of the above equations, the operations are those of the Heyting algebra $H$. Sendlewski's construction can give an intuitive meaning of Vakarelov's construction for Nelson algebras.

We now move to Pagliani's constructions. Let $U$ be a set and $E$ be an equivalence relation. Approximations are then defined in terms of an *indiscernibility space*, that is, a relational structure $(U, E)$ such that $E$ is an equivalence relation on $U$.

For a subset of $X$ of $U$, we define the lower approximation $X_E$ of $X$ which consists of all elements whose $E$-class is included in $X$ and the upper approximation $X^E$ of $X$ which is the set of the elements whose $E$-class has non-empty intersection with $X$.

Therefore, $X_E$ can be viewed as the set of elements which *certainly* belong to $X$ and $X^E$ is the set of objects that *possibly* are in $X$, when elements are observed through the knowledge synthesized by $E$.

In this setting, we can use arbitrary binary relations instead of equivalence relations. For this purpose, we introduce approximations $(\cdot))_R$ and $(\cdot)^R$, where $R$ is reflexive. They can be regarded as "real" lower and upper approximation operators.

**Definition 4.2** Let $R$ be a reflexive relation on $U$ and $X \subseteq U$. The set $R(X) = \{y \in U \mid xRy$ for some $x \in X\}$ is the *R-neghbourhood* of $X$. If $X = \{a\}$, then we write $R(a)$ instead of $R(\{a\})$. The approximations are defined as $X_R = \{x \in U \mid R(x) \subseteq X\}$ and $X^R = \{x \in U \mid R(X) \cap \neq \emptyset\}$. A set $X \subseteq U$ is called *R-closed* if $R(X) = X$ and an element $x \in U$ is *R-closed*, if its singleton set $\{x\}$ is $R$-closed. The set of $R$-closed points is denoted by $S$.

The rough set of $X$ is the equivalence class of all $Y \subseteq U$ such that $Y_E = X_E$ and $Y^E = X^E$. Since each rough set is uniquely determined by the approximation pair, it is possible to represent the rough set of $X$ as $(X_E, X^E)$ or $(X_E, -X^E)$. We call the former *increasing representation* and the latter *disjoint representation*.

These representations induce the sets:

$$IRS_E(U) = \{(X_E, X^E) \mid X \subseteq U\}$$

and

$$DRS_E(U) = \{(X_E, -X^E) \mid X \subseteq U\},$$

respectively. The set $IRS_E(U)$ can be ordered componentwise

$$(X_E, X^E) \leq (Y_E, Y^E) \Leftrightarrow X_E \subseteq Y_E \text{ and } X^E \subseteq Y^E,$$

and $DRS_E(U)$ is ordered by reversing the order for the second components of the pairs:

$$\begin{aligned}(X_E. - X^E) \leq (Y_E, -Y^E) &\Leftrightarrow X_E \subseteq Y_E \text{and} - X^E \supseteq -Y^E \\ &\Leftrightarrow X_E \subseteq Y_E \text{and}\, X^E \subseteq Y^E\end{aligned}$$

Therefore, $IRS_E(U)$ and $DRS_E(U)$ are order-isomorphic, and they form completely distributive lattices.

Every Boolean lattice $B$, where $x'$ denotes the complement of $x \in B$, is a Heyting algebra such that $x \Rightarrow y = x' \vee y$ for $x, y \in B$. An element $x \in B$ is dense only if $x' = 0$, that is, $x = 1$.

Because it is known that on a Boolean lattice each lattice-congruence is such that the quotient lattice is a Boolean lattice, also the congruence $\cong_S$ on $\mathscr{B}_E(U)$, defined by $X \cong_S Y$, if $X \cap S = Y \cap S$, is Boolean when $\mathscr{B}_E(U)$ is interpreted as a Heyting algebra.

Pagliani [12] showed that the disjoint representation of rough sets can be characterized as

$$DRS_E(U) = \{(A, B) \in \mathscr{B}_E(U)^2 \mid A \cap B = \emptyset \text{ and } A \cup B \cong_S U\}$$

Thus, $DRS_E(U)$ coincides with the Nelson lattice $N_{\cong_S}(\mathscr{B}_E(U))$. Since $\mathscr{B}_E(U)$ is a Boolean lattice, $N_{\cong_S}(\mathscr{B}_E(U))$ is a semi-simple Nelson algebra. Consequently, we obtain that rough sets defined by equivalences determined also regular double Stone algebras and three-valued Lukasiewicz algebras.

Järvinen et al. [13] further extended Pagliani's results for the so-called *effective lattice*, which expands Nelson algebras with a modal operator capable of expressing the classical truth. The feature is of special importance for the problems in computer science, AI, and natural language semantics.

## 4.5 Three-Valued Logics and Rough Sets

Since rough sets are used to describe incompleteness of information, it is natural to consider a foundation for rough set theory based on *three-valued logic*. This line of study can be found in Akama and Murai [18] and Avron and Konikowska [19]. In this section, we review briefly Avron and Konikowska's approach.

Let $K = (U, \mathbf{R})$, $R \in \mathbf{R}$ and $X \subseteq U$. Then, as presented in Chap. 2, in rough set theory, one can classify the three regions, i.e., $R$-positive region $POS_R(X)$, $R$-negative region $NEG_R(X)$ and $R$-boundary region $BN_R(X)$.

The elements of $POS_R(X)$ *certainly belong* to $X$, the elements of $NEG_R(X)$ *certainly do not belong* to $X$, and we *cannot tell if* the elements of $BN_R(X)$ belong to $X$ or not. Because the classification can be related to three-valued language, the idea led Avron and Konikowska to a three-valued logic approach to rough sets.

They use a simple three-valued logic $\mathscr{L}_{rs}$, whose formulas are all expressions of the form $Ax$ where $A$ is an expression representing a subset of $U$ and $x$ is a variable representing an object in $U$.

The semantics of $\mathscr{L}_{rs}$ uses logical values in $\mathscr{T} = \{t, f, u\}$, where

- $t$ represents the classical value true,
- $f$ represents the classical value false,
- $u$ represents a non-classical value unknown.

As designated values we can take either $\{u, t\}$-obtaining a weaker logic-or only $\{t\}$, obtaining a strong logic.

The truth-values of formula $\mathscr{L}_{rs}$ with respect to a knowledge base $k = \langle U, \mathbf{R} \rangle$, a relation $R \in \mathbf{R}$, an interpretation $| \cdot |$ of set expressions, and a valuation $v$ of object variables, as follows:

$$(1) \|Ax\|_v = \begin{cases} t \text{ if } v(x) \in POS_R(| A |) \\ f \text{ if } v(x) \in NEG_R(| A |) \\ u \text{ if } v(x) \in BN_R(| A |) \end{cases}$$

Unfortunately, the logic $\mathscr{L}_{rs}$ has a major drawback, namely its semantics is not decompositional. This follows from the fact that the lower and upper approximations of a set obey the rules:

$$\begin{array}{lll} & \overline{R}(A \cup B) = \overline{R}A \cup \overline{R}B & \underline{R}(A \cup B) \supseteq \underline{R}A \cup \underline{R}B \\ (2) & \underline{R}(A \cap B) = \underline{R}A \cap \underline{R}B & \overline{R}(A \cap B) \subseteq \overline{R}A \cup \overline{R}B \\ & \underline{R}(-A) = -\overline{R}A & \overline{R}(-A) = -\underline{R}A \end{array}$$

where the inclusions cannot be in general replaced by equalities.

Clearly, these inclusions imply that the values of $(A \cup B)x$ and $(A \cap B)x$ are not always uniquely determined by the values of $Ax$ and $Bx$, which is exactly the factor that makes the semantics of $\mathscr{L}_{rd}$ non-decompositional. Namely, we have:

If $\|Ax\| = u$ and $\|Bx\| = u$, then
$\|(A \cup B)x\| = \{u, t\}$ and $\|(A \cap B)x\| = \{f, u\}$

and we cannot in general say in advance which of the respective two values will be assigned by the interpretation to the considered two formulas.

The ordinary logical matrix cannot provide a semantics of $\mathscr{L}_{rs}$. To give a proper semantics, we need to use a *non-deterministic logical matrix* (Nmatrix) [20], which is a generalization of an ordinary matrix modelling non-determinism, with interpretations of logical connectives returning sets of logical values instead of single values.

**Definition 4.3** A non-deterministic matrix (Nmatrix) for a propositional language $L$ is a tuple $\mathscr{M} = (\mathscr{T}, \mathscr{D}, \mathscr{O})$, where:

- $\mathscr{T}$ is a non-empty set of truth-values.
- $\emptyset \subset \mathscr{D} \subseteq \mathscr{T}$ is the set of designated values.
- For every $n$-ary connective $\diamond$ of $L$, $\mathscr{O}$ includes a corresponding $n$-ary function $\tilde{\diamond}$ from $\mathscr{T}^n$ to $2^{\mathscr{T}} - \{\emptyset\}$.

Let $W$ be the set of well-formed formulas of $L$. A (legal) valuation in an Nmatrix $\mathscr{M}$ is a function $v : W \to \mathscr{T}$ that satisfies the following condition:

$$v(\diamond(\psi_1, ..., \psi_n)) \in \tilde{\diamond}\, (\psi_1, ..., \psi_n))$$

for every $n$-ary connective $\diamond$ of $L$ and any $\psi_1, ..., \psi_n \in W$.
Let $\mathscr{V}_M$ dente the set of all valuations in the Nmatrix $\mathscr{M}$. The notions of satisfaction under a valuation, validity and consequence relation are defined as follows:

- A formulas $\phi \in W$ is satisfied by a valuation $v \in \mathscr{V}_M$, denoted $v \models \phi$, if $v(\phi) \in \mathscr{D}$.
- A sequent $\Sigma = \Gamma \Rightarrow \Delta$ is satisfied by a valuation $v \in \mathscr{V}_M$, denoted $v \models \Sigma$, iff either $v$ does not satisfy some formula in $\Gamma$ or $v$ satisfies some formula in $\Delta$.
- A sequent $\Sigma$ is valid, denoted $\models \Sigma$, if it is satisfied by all valuations $v \in \mathscr{V}_M$.
- The consequence relation on $W$ defined by $\mathscr{M}$ is the relation $\vdash_{\mathscr{M}}$ on sets of formulas in $W$ such that, for any $T, S \subseteq W, T \vdash_{\mathscr{M}} S$ iff there exist finite sets $\Gamma \subseteq T, \Delta \subseteq S$ such that the sequent $\Gamma \Rightarrow \Delta$ is valid.

We can define a three-valued logic for modelling rough sets with semantics based on Nmatrix; see Avron and Lev [20]. Avron and Konikowska used a simple predicate language with atomic formulas expressing membership of objects in sets.

**Definition 4.4** The alphabet of a simple predicate language $L_P$ contains:

- a set $\mathbf{P}_n$ of $n$-ary predicate symbols for $n = 0, 1, 2, \ldots$ .
- a set $\mathbf{O}_n^k$ of symbols for $k$-ary operations on $n$-ary predicates for $n, k = 0, 1, 2, \ldots$ .
- a set of $\mathbf{V}$ of individual variables.

The set $\mathbf{E}_n$ of predicate expressions of arity $n$ is the least set such that:

- $\mathbf{P}_n \subseteq \mathbf{E}_n$.
- if $\diamond \in \mathbf{O}_n^k$ and $e_1, ..., e_k \in \mathbf{E}_n$, then $\diamond(e_1, ..., e_k) \in \mathbf{E}_n$.

- The consequence relation on $W$ defined by $\mathcal{M}$ is the relation $\vdash_{\mathcal{M}}$ on sets of formulas in $W$ such that, for any $T, S \subseteq W$, $T \vdash_{\mathcal{M}} S$ iff there exist finite sets $\Gamma \subseteq T$, $\Delta \subseteq S$ such that the sequent $\Gamma \Rightarrow \Delta$ is valid.

The set $W$ of well-formed formulas of $L_P$ consists of all expressions of the form
$e(x_1, ..., x_n)$
where $n \geq 0$, $e \in \mathbf{E}_n$ and $x_1, ..., x_n \in \mathbf{V}$.

The semantics of $L_P$ is defined based on an Nmatrix for $L_P$ and a structure for $L_P$.

**Definition 4.5** An Nmatrix for $L_P$ is a non-deterministic matrix $\mathcal{M} = (\mathcal{T}, \mathcal{D}, \mathcal{O})$ with $\mathcal{O}$ containing an interpretation $\tilde{\diamond}$: $\mathcal{T}^k \rightarrow 2^{\mathcal{T}}/\{\emptyset\}$ for every $k$-ary operation $\diamond$ on $n$-ary predicate in $\mathbf{O}_n^k$, $n \geq 0$.

**Definition 4.6** A $\mathcal{T}$-structure for $L_P$ is a pair $\mathbf{M} = (X, | \cdot |)$, where

- $X$ is a non-empty set.
- $| \cdot |$ is an interpretation of predicate symbol, with $| \ p \ |$: $X^n \rightarrow \mathcal{T}$ for any $p \in P_n, n \geq 0$.

To define the interpretation of $L_P$ in a given structure, we use a non-deterministic matrix for interpreting the operators of that language.

**Definition 4.7** Let $\mathcal{M} = (\mathcal{T}, \mathcal{D}, \mathcal{O})$ be an Nmatrix for $L_P$, and let $\mathcal{M} = (X, | \cdot |)$ be a $\mathcal{T}$-structure for $\mathcal{L}_P$.

An interpretation of $L_P$ under the Nmatrix $\mathcal{M} = (\mathcal{T}, \mathcal{D}, \mathcal{O})$ in the structure $\mathcal{M} = (X, | \cdot |)$ for a valuation $c : V \rightarrow X$ is a function $\| \cdot \|_v^M : W \rightarrow \mathcal{T}$ such that:

- $\|p(x_1, ..., x_n)\| =| \ p \ | (v(x_1), ..., v(x_n))$ for any $p \in \mathbf{P}_n, n \geq 0$.
- $\| \diamond (e_1, ..., e_n)(x_1, ..., x_n)\|_v^M \in \tilde{\diamond} (\|e_1(x_1, ..., x_n))\|_v^M, ..., \|e_k(x_1, ..., x_n)\|_v^M)$

for any $k$-ary operation on $n$-ary predicates $\diamond \in \mathbf{O}_n^k$, and $n$-ary predicate expressions $e_1, ..., e_n \in \mathbf{E}_n$, and any individual variables $x_i \in \mathbf{V}$, $i = 1, ..., n$.

To simplify notation, in what follows we will drop the decoration on $\| \cdot \|$ symbol.

Avron and Konikowska define propositional rough set logic. The predicate language $L_{RS}$ for describing rough sets uses only unary predicate symbols representing sets, object variables, and the symbols $-, \cup, \cap$, i.e.,

- $\mathbf{P}_1 = \{P, Q, R, ...\}$, $\mathbf{P}_n = \emptyset$ for $n \neq 1$
- $\mathbf{O}_1^1 = \{-\}$, $\mathbf{O}_1^2 = \{\cup, \cap\}$, $\mathbf{O}_n^k = \emptyset$ otherwise.

Thus, the set $W_{RS}$ of well-formed formulas of $L_{RS}$ contains all expressions of the form $Ax$, where $A$ is a unary predicate expression representing a set, built of the predicate symbols in $\mathbf{P}_1$ and using the operation symbols $-, \cup, \cap$, while $x \in \mathbf{V}$ is an individual variable.

The semantics of $L_{RS}$ is given by the Nmatrix $\mathcal{M}_{RS} = (\mathcal{T}, \mathcal{D}, \mathcal{O})$, where $\mathcal{T} = \{f, u, t\}$, $\mathcal{D} = \{t\}$, and $-, \cup, \cap$ are interpreted as set-theoretic operations on rough sets, with their semantics given by:

| $\tilde{-}$ | $f$ | $u$ | $t$ |
|---|---|---|---|
| | $t$ | $u$ | $f$ |

| $\tilde{\cup}$ | $f$ | $u$ | $t$ |
|---|---|---|---|
| $f$ | $f$ | $u$ | $t$ |
| $u$ | $u$ | $\{u, t\}$ | $t$ |
| $t$ | $t$ | $t$ | $t$ |

| $\tilde{\cap}$ | $f$ | $u$ | $t$ |
|---|---|---|---|
| $f$ | $f$ | $f$ | $f$ |
| $u$ | $f$ | $\{f, u\}$ | $u$ |
| $t$ | $f$ | $u$ | $t$ |

where $f$, $u$ and $t$ stand for the appropriate singleton sets.

By the rules (2) governing the interplay between the operations of lower and upper approximations and set-theoretic operations in the rough sets framework, the interpretation of the latter operations in the Nmatrix $\mathscr{M}_{RS}$ corresponds to the intended interpretation (1) of the $Ax$ type of formulas in that framework.

Note that complement is deterministic, while union and intersection are non-deterministic. In fact, the results of the operation on two undefined arguments for union and intersection are non-deterministic.

Now, we present a three-valued rough set logic $\mathscr{L}_{RS}$ defined by the language $L_{RS}$ with the semantics given by $\mathscr{M}_{RS}$, in which the implication is defined as:

$A \rightarrow B = \neg A \vee B$

where $\neg$, $\vee$ correspond to $-$, $\cup$ in $\mathscr{M}_{RS}$.

A propositional rough set logic $\mathscr{L}^I_{RS}$ is defined by propositional variables in $\mathbf{P} = \{p, q, r, ...\}$ and connectives $\neg$, $\rightarrow$. The formulas of $\mathscr{L}^I_{RS}$ are denoted by $A$, $B$, $C$, ..., and the set of all well-formed formulas by $W_I$.

The Nmatrix corresponding to $\mathscr{L}^I_{RS}$ is $\mathscr{M}^I_{RS} = (\mathscr{T}, \mathscr{D}, \mathscr{O}_I)$, where $\mathscr{T}$, $\mathscr{D}$ are the same as before, and $\mathscr{O}_I$ contains the interpretations of $\neg$ and $\rightarrow$ defined by the following:

| $\tilde{\neg}$ | $f$ | $u$ | $t$ |
|---|---|---|---|
| | $t$ | $u$ | $f$ |

| $\tilde{\rightarrow}$ | $f$ | $u$ | $t$ |
|---|---|---|---|
| $f$ | $t$ | $t$ | $t$ |
| $u$ | $u$ | $\{u, t\}$ | $t$ |
| $t$ | $f$ | $u$ | $t$ |

Observe that the logic based on $\mathscr{M}^I_{RS}$ can be seen as a "common denominator" of three-valued logics of Kleene and Lukasiewicz. This is because these two famous three-valued logics are distinguished by the non-deterministic part of the above truth-value table for implication.

Avron and Konikowska described a sequent calculus for the logic generated by $\mathscr{M}^I_{RS}$ with a completeness result. In addition, they discuss the relations of Kleene's and Lukasiewicz's three-valued logics, claiming that $\mathscr{L}^I_{RS}$ is the common part of these two three-valued logics.

A *sequent calculus* was proposed by Gentzen [21] for classical and intuitionistic logic. He also developed *natural deduction*. Sequent calculus is considered to be more convenient than natural deduction for some technical reasons.

$\mathscr{L}^{I}_{RS}$ is based on $\mathscr{M}^{I}_{RS}$. The logic has no tautologies, but only valid entailment, represented by valid sequents. Thus, the appropriate formulation is a sequent calculus. **IRS** is the sequent calculus over $\mathscr{L}^{I}_{RS}$.

Let $\Gamma$, $\Delta$ be sets of formulas, and $A$, $B$ be formulas. A *sequent* is an expression of the form $\Gamma \Rightarrow \Delta$, where $\Gamma$ is called the *antecedent* and $\Delta$ is called the *succedent*, respectively. Note that formulas in the antecedent are interpreted conjunctively and formulas in the succedent are interpreted disjunctively.

A sequent calculus is formalized by sets of *axioms* and *rules*. Rules are divided into *structural rules* and *logical rules*.

**IRS**
**Axioms**
(A1) $A \Rightarrow A$,
(A2) $\neg A, A \Rightarrow$
**Inference Rules**

Weakening on both sides, together with the following rules:

$$(Weakening)\frac{\Gamma \Rightarrow \Delta}{\Gamma' \Rightarrow \Delta'}$$

In$(Weakening)$, $\Gamma \subseteq \Gamma', \Delta \subseteq \Delta'$.

$$(\neg\neg \Rightarrow)\frac{\Gamma, A \Rightarrow \Delta}{\Gamma, \neg\neg A \Rightarrow \Delta} \qquad (\Rightarrow \neg\neg)\frac{\Gamma \Rightarrow \Delta, A}{\Gamma \Rightarrow \Delta, \neg\neg A}$$

$$(\Rightarrow\rightarrow I)\frac{\Gamma \Rightarrow \Delta, \neg A}{\Gamma \Rightarrow \Delta, A \rightarrow B} \qquad (\Rightarrow\rightarrow II)\frac{\Gamma \Rightarrow \Delta, B}{\Gamma \Rightarrow \Delta, A \rightarrow B}$$

$$(\rightarrow\Rightarrow I)\frac{\Gamma \Rightarrow \Delta, A \quad \Gamma, B \Rightarrow \Delta}{\Gamma, A \rightarrow B \Rightarrow \Delta} \qquad (\rightarrow\Rightarrow II)\frac{\Gamma, \neg A \Rightarrow \Delta \quad \Gamma \Rightarrow \Delta, \neg B}{\Gamma, A \rightarrow B \Rightarrow \Delta}$$

$$(\Rightarrow \neg \rightarrow)\frac{\Gamma \Rightarrow \Delta, A \quad \Gamma \Rightarrow \Delta, \neg B}{\Gamma \Rightarrow \Delta, \neg(A \rightarrow B)}$$

$$(\neg \rightarrow\Rightarrow I)\frac{\Gamma, A \Rightarrow \Delta}{\Gamma, \neg(A \rightarrow B) \Rightarrow \Delta} \qquad (\neg \rightarrow\Rightarrow II)\frac{\Gamma, \neg B \Rightarrow \Delta}{\Gamma, \neg(A \rightarrow B) \Rightarrow \Delta}$$

It is possible to obtain a set of rules which is more compact than the above. Namely, rules $(\Rightarrow\rightarrow I)$ and $(\Rightarrow\rightarrow II)$ can be combined to:

$$\frac{\Gamma \Rightarrow \Delta, \neg A, B}{\Gamma \Rightarrow \Delta, A \rightarrow B}$$

while rules $(\rightarrow\Rightarrow I)$ and $(\rightarrow\Rightarrow II)$ can be combined to:

$$\frac{\Gamma, \neg A \Rightarrow \Delta \quad \Gamma, B \Rightarrow \Delta \quad \Gamma \Rightarrow \Delta, A, \neg B}{\Gamma, A \rightarrow B \Rightarrow \Delta}$$

and, finally,rules $(\neg \rightarrow\Rightarrow I)$ and $(\neg \rightarrow\Rightarrow II)$ can be combined to:

$$\frac{\Gamma, A, \neg B \Rightarrow \Delta}{\Gamma, \neg(A \rightarrow B) \Rightarrow \Delta}.$$

As we can see, there is a clear trade-off between the size of the set of rules and complexity of the individual rules. So the choice of a particular option should depend on the intended application.

**Lemma 4.2** *The following hold:*

1. *The axioms of the system* **IRS** *are valid.*
2. *For any inference rule $r$ of* **IRS** *and any valuation $v$, if $v$ satisfies all the premises of $r$ then $v$ satisfies the conclusion of $r$.*

As a corollary of Lemma 4.2, the inference rules of **IRS** are sound, i.e., they preserve the validity of sequents.

**Theorem 4.3** *The sequent calculus* **IRS** *is sound and complete for* $\vdash_{\mathscr{M}^{1}_{RS}}$.

As disjunction and conjunction can be represented by negation and implication according to the relationships:

$$A \vee B =_{\text{def}} \neg A \rightarrow B$$
$$A \wedge B =_{\text{def}} \neg(A \rightarrow \neg B),$$

they can be treated as derived operations in our language.

From the above representation and **IRS**, we can derive the following sequent rule for conjunction and disjunction:

$$\frac{\Gamma \Rightarrow \Delta, A}{\Gamma \Rightarrow \Delta, A \vee B} \qquad \frac{\Gamma \Rightarrow \Delta, B}{\Gamma \Rightarrow \Delta, A \vee B}$$

$$\frac{\Gamma \Rightarrow \Delta, \neg A \quad \Gamma \Rightarrow \Delta, \neg B}{\Gamma \Rightarrow \Delta, \neg(A \vee B)}$$

$$\frac{\Gamma, \neg A \Rightarrow \Delta}{\Gamma. \neg(A \vee B) \Rightarrow \Delta} \qquad \frac{\Gamma, \neg B \Rightarrow \Delta}{\Gamma. \neg(A \vee B) \Rightarrow \Delta}$$

$$\frac{\Gamma, A \Rightarrow \Delta \quad \Gamma \Rightarrow \Delta, \neg B}{\Gamma, A \vee B \Rightarrow \Delta} \qquad \frac{\Gamma, B \Rightarrow \Delta \quad \Gamma \Rightarrow \Delta, \neg A}{\Gamma, A \vee B \Rightarrow \Delta}$$

$$\frac{\Gamma \Rightarrow \Delta, A \quad \Gamma \Rightarrow \Delta, B}{\Gamma \Rightarrow \Delta, A \wedge B}$$

$$\frac{\Gamma \Rightarrow \Delta, \neg A}{\Gamma \Rightarrow \Delta, \neg(A \wedge B)} \qquad \frac{\Gamma \Rightarrow \Delta, \neg B}{\Gamma \Rightarrow \Delta, \neg(A \wedge B)}$$

$$\frac{\Gamma, A \Rightarrow \Delta}{\Gamma A \wedge B \Rightarrow \Delta} \qquad \frac{\Gamma, B \Rightarrow \Delta}{\Gamma A \wedge B \Rightarrow \Delta}$$

$$\frac{\Gamma, \neg A \Rightarrow \Delta \quad \Gamma \Rightarrow \Delta, B}{\Gamma, \neg(A \wedge B) \Rightarrow \Delta} \qquad \frac{\Gamma, \neg B \Rightarrow \Delta \quad \Gamma \Rightarrow \Delta, A}{\Gamma, \neg(A \wedge B) \Rightarrow \Delta}$$

The above system, with the negation rule of **IRS**, can be used for reasoning in the rough set logic $\mathscr{L}_{RS}$. If we first apply a translation $\tau$ from the language of $\mathscr{L}_{RS}$ to the language $\mathscr{L}'_{RS}$ of atomic predicate expressions combined with $\neg, \vee, \cap$ such that:

$\tau((A \cup B)x) = \tau(Ax) \vee \tau(Bx)$
$\tau((A \cap B)x) = \tau(Ax) \wedge \tau(Bx)$
$\tau((-A)x) = \neg\tau(Ax)$

then the system mentioned above, together with the substitution principle, is complete for $\mathscr{L}'_{RS}$, and so also for $\mathscr{L}_{RS}$.

Avron and Konikowska further discussed the relations with Kleene and Lukasiewicz three-valued logics, by showing $\mathscr{L}^I_{RS}$ as the common part of these three-valued logics.

Let us denote by $\mathscr{L}_K$ and $\mathscr{L}_L$ Kleene's and Lukasiewicz's three-valued logics, respectively. Recall that $\mathscr{L}^I_{RS}$ is given by the Nmatrix $\mathscr{M}^I_{RS} = (\mathscr{T}, \mathscr{D}, \{\tilde{\neg}, \tilde{\rightarrow}\})$, where the following are satisfied.

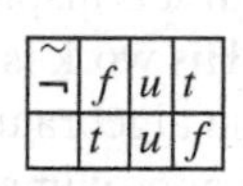

| $\tilde{\neg}$ | $f$ | $u$ | $t$ |
|---|---|---|---|
| | $t$ | $u$ | $f$ |

| $\tilde{\rightarrow}$ | $f$ | $u$ | $t$ |
|---|---|---|---|
| $f$ | $t$ | $t$ | $t$ |
| $u$ | $u$ | $\{u, t\}$ | $t$ |
| $t$ | $f$ | $u$ | $t$ |

Now, using the same notational convention, we can say that the $\{\neg, \rightarrow\}$ version of Kleene and Lukasiewicz logics are given by ordinary mattrices $\mathscr{M}_L$ and $\mathscr{M}_L$ with $\mathscr{T}$, $\mathscr{D}$ and the common interpretation $\tilde{\neg}$ of negation as in the Nmatrix $\mathscr{M}^I_{RS}$.

But, different interpretations of implication, given, respectively, by $\rightarrow_K$ and $\rightarrow_L$ are defined below:

| $\rightarrow_L$ | $f$ | $u$ | $t$ |
|---|---|---|---|
| $f$ | $t$ | $t$ | $t$ |
| $u$ | $u$ | $t$ | $t$ |
| $t$ | $f$ | $u$ | $t$ |

| $\rightarrow_K$ | $f$ | $u$ | $t$ |
|---|---|---|---|
| $f$ | $t$ | $t$ | $t$ |
| $u$ | $u$ | $u$ | $t$ |
| $t$ | $f$ | $u$ | $t$ |

From these three mattrices, $\mathscr{M}_K$ and $\mathscr{M}_L$ are included in the Nmatirix $\mathscr{M}^I_{RS}$, and represent its two different "determinizations". Then, we have the following:

**Lemma 4.3** *The system* **IRS** *is sound for both Kleene and Lukasiewicz logics.*

To make **IRS** also complete for $\mathscr{L}_K$ and $\mathscr{L}_L$, it suffices to add just one sequent rule for each logic:

**Theorem 4.4** *Let* $(K)$ *and* $(L)$ *be the two following sequent rules:*

$$(K)\frac{\Gamma, \neg A \Rightarrow \Delta \quad \Gamma, B \Rightarrow \Delta}{\Gamma, A \rightarrow B \Rightarrow \Delta} \qquad (L)\frac{\Gamma, A \Rightarrow \Delta \quad \Gamma, \neg B \Rightarrow \Delta}{\Gamma \Rightarrow \Delta, A \rightarrow B}$$

*Then, we have the following:*

1. *The system* $\mathbf{IRS}^K$ *obtained by adding rule* $(K)$ *to* **IRS** *is sound and complete for Kleene logic.*
2. *The system* $\mathbf{IRS}^L$ *obtained by adding rule* $(L)$ *to* **IRS** *is sound and complete for Lukaisewicz logic.*

The three-valued logic approach appeared to open a new possibility of developing rough set theory. If we consider other three-valued logics, we could have a new type of rough set theory. In addition, we should explore other many-valued logics, e.g., four-valued logic for this purpose.

## 4.6 Rough Set Logics

It is necessary to advance a logical system based on rough sets for practical applications. Such a logic is called a *rough set logic*. The first approach to a rough set logic was established by Düntsch [22] in 1997. Here, we quickly review his system and related ones.

Düntsch developed a propositional logic for rough sets inspired by the topological construction of rough sets using Boolean algebras. His work is based on the fact that the collection of all subsets of a set forms a Boolean algebra under the set-theoretic operation, and that the collection of rough sets of an approximation space is a regular double Stone algebra. Thus, we can assume that regular double Stone algebras can serve as a semantics for a logic for rough sets.

To understand his logic, we need some concepts. A *double Stone algebra* $DSA$ is denoted by $\langle L, +, \cdot, *.+, 0, 1\rangle$ with the type $\langle 2, 2, 1, 1, 0, 0\rangle$ satisfying the following conditions:

(1) $\langle L, +, \cdot, 0, 1\rangle$ is a bounded distributed lattice.
(2) $x^*$ is the pseudocomplement of $x$, i.e., $y \leq x^* \Leftrightarrow y \cdot x = 0$.
(3) $x^+$ is the dual pseudocomplement of $x$, i.e., $y \geq x^+ \Leftrightarrow y + x = 1$.
(4) $x^* + x^{**} = 1, x^+ \cdot x^{++} = 0$

$DSA$ is called *regular* if it satisfies the additional condition: $x \cdot x^+ \leq x + x^*$. Let $B$ be a Boolean algebra, $F$ be a filter on $B$, and $\langle B, F\rangle = \{\langle a, b\rangle \mid a, b \in B, a \leq b, a + (-b) \in F\}$.

We define the following operations on $\langle B, F\rangle$ as follows:

$$\langle a, b\rangle + \langle c, d\rangle = \langle a + c, b + d\rangle,$$
$$\langle a, b\rangle \cdot \langle c, d\rangle = \langle a \cdot c, b \cdot d\rangle,$$
$$\langle a, b\rangle^* = \langle -b, -b\rangle,$$
$$\langle a, b\rangle^+ = \langle -a, -a\rangle.$$

If $\langle U, R\rangle$ is an approximation space, the classes of $R$ can be viewed as a complete subalgebra of the Boolean algebra $B(U)$. Conversely, any atomic complete subalgebra $B$ of $B(U)$ yields an equivalence relation $R$ on $U$ by the relation: $xRy \Leftrightarrow x$ and $y$ are contained in the same atom of $B$, and this correspondence is bijective.

If $\{a\} \in B$, then for every $X \subseteq U$ we have: If $a \in \underline{R}X$, then $a \in X$, and the rough sets of the corresponding approximation space are the elements of the regular double Stone algebra $\langle B, F\rangle$, where $F$ is the filter of $B$ which is generated by the union of the singleton elements of $B$.

Based on the construction of regular double Stone algebras, Düntsch proposed a propositional rough set logic $RSL$. The language $\mathscr{L}$ of $RSL$ has two binary connectives $\wedge$ (conjunction), $\vee$ (disjunction), two unary connectives $^*, ^+$ for two types of negation, and the logical constant $\top$ for truth.

Let $P$ be a non-empty set of propositional variables. Then, the set **Fml** of formulas with the logical operators constitutes an absolutely free algebra with a type $\langle 2, 2, 1, 1, 0 \rangle$. Let $W$ be a set and $B(W)$ be a Boolean algebra based on $W$.

Then, a *model* $M$ of $L$ is seen as a pair $(W, v)$, where $v : P \rightarrow B(W) \times B(W)$ is the *valuation function* for all $p \in P$ satisfying: if $v(p) = \langle A, B \rangle$, then $A \subseteq B$. Here, $v(p) = \langle A, B \rangle$ states that $p$ holds at all states of $A$ and does not hold at any state outside $B$.

Düntsch relates the valuation to Lukasiewicz's three-valued logic by the following construction. For each $p \in P$, let $v_p : W \rightarrow \mathbf{3} = \{0, \frac{1}{2}, 1\}$. $v : P \rightarrow B(W) \times B(W)$ is defined as follows: $v(p) = \langle \{w \in W : v_p(w) = 1\}, \{w \in W : v_p(w) \neq 0\} \rangle$. In addition, Düntsch connected the valuation and rough sets as follows:

$v_p(w) = 1$ if $w \in A$,
$v_p(w) = \frac{1}{2}$ if $w \in B \setminus A$,
$v_p(w) = 0$ otherwise.

Given a model $M = (W, v)$, the *meaning function* $\mathtt{mng} : \mathbf{Fml} \rightarrow \mathtt{B}(W) \times \mathtt{B}(W)$ is defined to give a valuation of arbitrary formulas in the following way:

$\mathtt{mng}(\top) = \langle W, W \rangle$,
$\mathtt{mnf}(\rightarrow p) = \langle W, W \rangle$,
$\mathtt{mng}(p) = v(p)$ for $p \in P$.
If $\mathtt{mng}(\phi) = \langle A, B \rangle$ and $\mathtt{mng}(\psi) = \langle C, D \rangle$, then
$\mathtt{mng}(\phi \wedge \psi) = \langle A \cap C, B \cap D \rangle$,
$\mathtt{mng}(\phi \vee \psi) = \langle A \cup C, B \cup D \rangle$,
$\mathtt{mng}(\phi^*) = \langle -B, -B \rangle$,
$\mathtt{mng}(\phi^+) = \langle -A, -A \rangle$.

Here, $-A$ denotes the complement of $A$ in $B(W)$. We can understand that the meaning function assigns the meaning to formulas.

The class of all models of $L$ is denoted by **Mod**. A formula $A$ *holds* in a model $M = \langle W, v \rangle$, written $M \models A$, if $\mathtt{mng}(A) = \langle W, W \rangle$. A set $\Gamma$ of sentences *entails* a formula $A$, written $\Gamma \vdash A$, if every model of $\Gamma$ is a model of $A$.

We can define additional operations on **Fml** by

$A \rightarrow B = A^* \vee B \vee (A^+ \wedge B^{**})$
$A \leftrightarrow B = (A \rightarrow B) \wedge (B \rightarrow A)$

Düntsch proved several technical results including completeness.

**Theorem 4.5** *If* $M = \langle W, c \rangle \in$ **Mod** *and* $\phi, \psi \in$ **Fml**, *then*

1. $M \models \phi \leftrightarrow \psi$ *iff* $\mathtt{mng}(\phi) = \mathtt{mng}(\psi)$
2. $M \models \rightarrow p \leftrightarrow \phi$ *iff* $M \models \phi$.

Theorem 4.5 states completeness, compactness, and Beth definability property of $RSL$.

**Theorem 4.6** *The following hold:*

1. *$RSL$ has a finitely complete and strongly sound Hilbert style axiom system.*
2. *$RSL$ has a compactness theorem.*
3. *$RSL$ does not have the Beth definability property.*

The implication in $RSL$ is interesting, but it has no intuitive appeal. It is thus promising to extend $RSL$ with another implication. Such extensions can be found in Akama et al. in [23, 24], which relate rough set logics to Heyting-Brouwer logic and its sub-logic.

As discussed in Chap. 3, Nelson algebras give rise to an algebraic interpretation of rough sets. Since Nelson algebras is an algebraic semantics for constructive logic with strong negation, it may be possible to employ constructive logic with strong negation as another rough set logic. The idea has been explored in Pagliani's approach to rough sets reviewed above.

Düntsch's rough set logic is a starting point of the work of logics for rough sets. His logic is three-valued, and it should be compared to the logics of Avron and Konikowska presented in the previous section.

## 4.7 Logics for Reasoning About Knowledge

A first approach to connect rough set theory and modal logic was due to Orlowska's series of papers [25–27]. She implicitly showed the relation of Pawlak's rough sets and modal logic S5. However, she did more than this. Here, we review her logics for reasoning about knowledge.

Orlowska proposed a logic with knowledge operators which are relative to indiscernibility relations associated with agents, with a semantics based on rough sets. Orlowska's approach has the following three intuitions:

---

1. Knowledge of an agent about a predicate $F$ can be reflected by the ability of the agent to classify objects as instances or non-instances of $F$.
2. Knowledge of an agent about a sentence $F$ can be reflected by the ability of the agent to classify states into those in which $F$ is true and those in which $F$ is false.
3. With each agent there is associated to an indiscernibility relation, and the agent decides membership of objects or states up to this indiscernibility. As a consequence, knowledge operators are relative to indiscernibility.

---

Let $U$ be a universe (of states or objects) and let $AGT$ be a set of agents. For each $a \in AGT$, let $ind(a) \subseteq U \times U$ be an indiscernibility relation corresponding to agent $a$.

For a set $A$ of agents, we define indiscernibility $ind(A)$ as follows:

$(s, t) \in ind(A)$ iff $(s, t) \in ind(a)$ for all $a \in A$
$ind(\emptyset) = U \times U$.

Indiscernibility relations are equivalence relation (reflexive, symmetric and transitive) or similarity relations (reflexive and symmetric). Below, we confine ourselves to equivalence relations.

We here state the following properties of indiscernibility relations.

**Proposition 4.1** *Indiscernibility relations satisfy the following:*

1. $ind(A \cup B) = ind(A) \cap ind(B)$
2. $(ind(A) \cup ind(B))^* \subseteq ind(A \cap B)$
3. $A \subseteq B$ *implies* $ind(B) \subseteq ind(A)$

*Here,* $(ind(A) \cup ind(B))^* = ind(\{A, B\}^*)$ *is the set of all finite sequences with elements from the set* $\{A, B\}$.

**Proposition 4.2** *The family* $\{ind(A)\}_{A \subseteq AGT}$ *is a lower semilattice in which* $ind(AGT)$ *is the zero element.*

Now, we assume a subset $X$ of the universe $U$ and an indiscernibility relation $ind(A)$ for a certain set of agents. Since agents recognize elements of $U$ up to $ind(A)$, they grasp $X$ within the limit of tolerance determined by lower and upper approximation of $X$.

The lower approximation $\underline{ind}(A)X$ of $X$ with respect to $ind(A)$ is the union of those equivalence classes determined by $ind(A)$ which are included in $X$. The upper approximation $\overline{ind}(A)X$ of $X$ with respect to $ind(A)$ is the union of those equivalence classes determined by $ind(A)$ which have an element in common with $X$.

For non-transitive indiscernibility relations the respective definitions of approximations are obtained by taking similarity classes instead of equivalence classes.

**Proposition 4.3** $\underline{ind}(A)X$ *and* $\overline{ind}(A)X$ *satisfy the following:*

1. $x \in \underline{ind}(A)X$ *iff for all* $t \in U$ *if* $(x, t) \in ind(A)$, *then* $t \in X$.
2. $x \in \overline{ind}(A)X$ *iff there is* $t \in U$ *such that* $(x, t) \in ind(A)$ *and* $t \in X$.

**Proposition 4.4** *The following relations hold.*

1. $\underline{ind}(A)\emptyset = \emptyset$, $\overline{ind}(A)U = U$
2. $\underline{ind}(\emptyset)X = \emptyset$ *for* $X \neq U$, $\underline{ind}(\emptyset)U = U$,
   $\overline{ind}(\emptyset)X = U$ *for* $X \neq \emptyset$, $\overline{ind}(\emptyset)\emptyset = \emptyset$,
3. $\underline{ind}(A)X \subseteq X \subseteq \overline{ind}(A)X$
4. $\underline{ind}(A)\underline{ind}(A)X = \underline{ind}(A)X$, $\underline{ind}(A)\overline{ind}(A)X = \overline{ind}(A)X$,
   $\overline{ind}(A)\overline{ind}(A)X = \overline{ind}(A)X$, $\overline{ind}(A)\underline{ind}(A)X = \underline{ind}(A)X$
5. $X \subseteq Y$ *implies* $\underline{ind}(A)X \subseteq \underline{ind}(A)Y$ *and* $\overline{ind}(A)X \subseteq \overline{ind}(A)Y$
6. $ind(A) \subseteq ind(B)$ *implies* $\underline{ind}(B)X \subseteq \underline{ind}(A)X$ *and* $\overline{ind}(A)X \subseteq \overline{ind}(B)X$ *for any* $X \subseteq U$.

**Proposition 4.5** *The following relations for complement hold.*

1. $\overline{ind}(A)X \cup \underline{ind}(A)(-X) = U$
2. $\underline{ind}(A)X \cap \overline{ind}(A)(-X) = \emptyset$
3. $-\overline{ind}(A)X = \underline{ind}(A)(-X)$
4. $-\underline{ind}(A)X = \overline{ind}(A)(-X)$,

**Proposition 4.6** *The following relations for union and intersection hold.*

1. $\underline{ind}(A)X \cup \underline{ind}(A)Y \subseteq \underline{ind}(A)(X \cup Y)$
2. $X \cap Y = \emptyset$ *implies* $\underline{ind}(A)(X \cup Y) = \underline{ind}(A)X \cup \underline{ind}(A)Y$
3. $\underline{ind}(A)(X \cap Y) = \underline{ind}(A)X \cap \underline{ind}(A)Y$
4. $\overline{ind}(A)(X \cup Y) = \overline{ind}(A)X \cap \overline{ind}(A)Y$
5. $\overline{ind}(A)(X \cap Y) \subseteq \overline{ind}(A)X \cap \overline{ind}(A)Y$

A set $X \subseteq U$ is said to be

| | |
|---|---|
| $A$-definable | iff $\underline{ind}(A)X = X = \overline{ind}(A)X$ or $X = \emptyset$ |
| roughly $A$-definable | iff $\underline{ind}(A)X \neq \emptyset$ and $\overline{ind}(A)X \neq U$ |
| internally $A$-definable | iff $\underline{ind}(A)X = \emptyset$ |
| externally $A$-definable | iff $\overline{ind}(A)X = U$ |
| totally $A$-non-definable | iff it is internally $A$-non-definable and externally $A$-non-definable |

We can define sets of $A$-positive, $A$-negative and $A$-bordeline instances of a set $X \subseteq U$.

$POS(A)X = \underline{ind}(A)X$,
$NEG(A)X = -\overline{ind}(A)X$,
$BOR(A)X = \overline{ind}(A)X - \underline{ind}(A)X$

Intuitivelly, if $s \in POS(A)X$, then in view of agents from $A$ element $s$ is a member of $X$. If $s \in NEG(A)X$, then in view of agents from $A$ element $s$ is not a member of $X$. $BOR(A)X$ is the range of uncertainty. Element $s \in BOR(A)X$ whenever agents from $A$ cannot decide whether $s$ is a member of $X$ or not.

For these sets, we have the following propositions:

**Proposition 4.7** *We have the following:*

1. $POS(A)X, NEG(A)X, BOR(A)X$ *are pairwise disjoint.*
2. $POS(A)X \cup NEG(A)X \cup BOR(A)X = U$.
3. $POS(A)X, NEG(A)X, BOR(A)X$ *are A-definable.*

**Proposition 4.8** *We have the following:*

1. $A \subseteq B$ *implies* $POS(A)X \subseteq POS(B)X, NEG(A)X \subseteq NEG(B)X, BOR(B)X \subseteq BOR(A)X$.
2. $ind(A) \subseteq ind(B)$ *implies* $POS(B)X \subseteq POS(A)X, NEG(B)X \subseteq NEG(A)X, BOR(A)X \subseteq BOR(B)$

**Proposition 4.9** *We have the following:*

1. $POS(A)X \subseteq X, NEG(A)X \subseteq -X$
2. $POS(A)\emptyset = \emptyset, NEG(A)U = \emptyset$
2. $POS(\emptyset)X = \emptyset$ *if* $X \neq U, POS(\emptyset)U = U$
4. $NEG(\emptyset)X = \emptyset$ *if* $X \neq \emptyset, NEG(\emptyset)\emptyset = U$
5. $X \subseteq Y$ *implies* $POS(A)X \subseteq POS(A)Y, NEG(A)Y \subseteq NEG(A)X.$

**Proposition 4.10** *We have the following:*

1. $POS(A)X \cup POS(A)Y \subseteq POS(A)(X \cup Y)$
2. *If* $X \cap Y = \emptyset$ *then* $POS(A)(X \cup Y) = POS(A)X \cup POS(B)Y$
3. $POS(A)(X \cap Y) = POS(A)X \cap POS(A)Y$
4. $NEG(A)(X \cup Y) = NEG(A)X \cap NEG(A)Y$
5. $NEG(A)X \cup NEG(A)Y \subseteq NEG(A)(X \cup Y)$
6. $NEG(A)(-X) = POS(A)X$

**Proposition 4.11** *We have the following:*

1. $BOR(A)(X \cup Y) \subseteq BOR(A)X \cup BOR(A)Y$
2. $X \cap Y = \emptyset$ *implies* $BOR(A)(X \cup Y) = BOR(A)X \cup BOR(A)Y$
3. $BOR(A)(X \cap Y) \subseteq BOR(A)X \cap BOR(A)Y$
4. $BOR(A)(-X) = BOR(A)X$

**Proposition 4.12** *We have the following:*

1. $POS(A)X \cup POS(B)X \subseteq POS(A \cup B)X$
2. $POS(A \cap B)X \subseteq POS(A)X \cap POS(B)X$
3. $NEG(A)X \cup NEG(B)X \subseteq NEG(A \cup B)X$
4. $NEG(A \cap B)X \subseteq NEG(A)X \cap NEG(B)X.$

**Proposition 4.13** *We have the following:*

1. $POS(A)POS(A)X = POS(A)X$
2. $POS(A)NEG(A)X = NEG(A)X$
3. $NEG(A)NEG(A)X = -NEG(A)X$
4. $NEG(A)POS(A)X = -POS(A)X$

We define a family of knowledge operators $K(A)$ for $A \in AGT$: $K(A)X = POS(A)X \cup NEG(A)X.$

Intuitively, $s \in K(A)X$ whenever $s$ can be decided by agents from $A$ to be $A$-positive or $A$-negative instance of $X$.

**Proposition 4.14** *We have the following:*

1. $K(A)\emptyset, K(A)U = U$
2. $K(\emptyset)X = \emptyset$ *if* $X \neq U$
3. $ind(A) \subseteq ind(B)$ *implies* $K(B)X \subseteq K(A)X$ *for all* $X \subseteq U$
4. $A \subseteq B$ *implies* $K(A)X \subseteq K(B)X$ *for all* $X \subseteq U$

5. *If $X$ is $A$-definable, then $K(A)X = U$*

We say that knowledge of agents $A$ about $X$ is:

| | |
|---|---|
| complete | iff $K(A)X = U$ |
| incomplete | iff $BOR(A)X \neq \emptyset$ |
| rough | iff $POS(A)X, NEG(A)X, BOR(A)X \neq \emptyset$ |
| pos-empty | iff $POS(A)X = \emptyset$ |
| neg-empty | iff $NEG(A)X = \emptyset$ |
| empty | iff it is pos-empty and neg-empty. |

If knowledge of $A$ about $X$ is complete, then $A$ can discern $X$ from its complement. Every $A$ has a complete knowledge about any $A$-definable set, in particular, about $\emptyset$ and $U$. The fact that knowledge of any agent about the whole universe is complete should not be considered to be a paradox.

A predicate whose extension equals $U$ provides a trivial, in a sense, information. In any particular example $U$ represents the set of "all things perceivable by agents". However, if $U$ consists of all formulas of the predicate calculus, and $X \subseteq U$ is the set of all the valid formulas, then clearly not every agent has the complete knowledge about $X$, although he has the complete knowledge about $U$.

Observe that $X \subseteq Y$ does not imply $K(A)X \subseteq K(A)Y$, and $K(A)X$ is not necessarily included in $X$. By these facts, we can avoid the well-known paradoxes of epistemic logic, where all the formulas known by anyone are valid, and every agent knows all the logical consequences of his knowledge.

**Proposition 4.15** *The following conditions are equivalent:*

1. *$K(A)X$ is complete.*
2. *$X$ is $A$-definable.*
3. *$BOR(A)X = \emptyset$.*
4. *$POS(A)X = -NEG(A)X$.*

It follows that if agents $A$ have complete knowledge about $X$, then they can tell $X$ from its complement.

**Proposition 4.16** *The following conditions are equivalent:*

1. *$K(A)X$ is rough.*
2. *$X$ is roughly $A$-definable.*
3. *$\emptyset \neq BOR(A)X \neq U$.*
4. *$POS(A)X \subseteq -NEG(A)X$.*

**Proposition 4.17** *The following conditions are equivalent:*

1. *$K(A)X$ is pos-empty.*
2. *$X$ is internally $A$-non-definable.*
3. *$K(A)X = POS(A)X$.*
4. *$BOR(A)X = -NEG(A)X$.*

**Proposition 4.18** *The following conditions are equivalent:*

1. $K(A)X$ *is neg-empty.*
2. $X$ *is externally A-non-definable.*
3. $K(A)X = POS(A)X$.
4. $BOR(A)X = -POS(A)X$.

**Proposition 4.19** *The following conditions are equivalent:*

1. $K(A)X$ *is empty.*
2. $X$ *is totally A-non-definable.*
3. $BOR(A)X = U$.

**Proposition 4.20** *The following hold:*

1. $K(A)X \subseteq K(B)X$ *for all* $X \subseteq U$ *implies* $ind(B) \subseteq ind(A)$.
2. $ind(A) = ind(B)$ *iff* $K(A)X = K(B)X$ *for all* $X \subseteq U$.
3. $ind(A) \subseteq ind(B)$ *implies* $K(B)X \subseteq POS(B)K(A)X$ *and* $POS(A)K(B)X \subseteq K(A)X$.
4. $ind(A)$ *is the identity on* $U$ *iff* $K(A)X$ *is complete for all* $X$.

**Proposition 4.21** *The following hold:*

1. $K(A)X = K(A)(-X)$
2. $K(A)K(A)X = U$
3. $K(A)X \cup K(B)X \subseteq K(A \cup B)X$
4. $K(A \cap B)X \subseteq K(A)X \cap K(B)X$.

Next, we discuss the independence of agents. A set $A$ of agents is said to be *dependent* iff there is $B \subset A$ such that $K(A)X = K(B)X$ for all $X \subseteq U$. A set $A$ is *independent* if it is not dependent.

**Proposition 4.22** *The following conditions are equivalent:*

1. $A$ *is independent.*
2. *For every* $B \subset A$ *there is* $X \subseteq U$ *such that* $K(B)X \subset K(A)X$.

**Proposition 4.23** *The following hold:*

1. *If* $A$ *is independent, then every of its subsets is independent.*
2. *If* $A$ *is dependent, then every of its supersets is dependent.*

**Proposition 4.24** *If* $AGT$ *is independent, then for any* $A, B \subseteq AGT$*, the following conditions are satisfied:*

1. $K(A)X \subseteq K(B)X$ *for all* $X$ *implies* $A \subseteq B$.
2. $ind(A \cap B) = (ind(A) \cup ind(B))^*$.

The intuitive meaning of independence of a set of agents is that if we drop some agents from the independent set, then knowledge of the group of the remaining agents is less than knowledge of the whole group.

Similarly, if a set is dependent, then some of its elements are superfluous, we can drop them without changing knowledge of the group.

We say that a set $B$ of agents is *superfluous* in a set $A$ iff for all $X \subseteq U$, we have $K(A - B)X = K(A)X$.

**Proposition 4.25** *The following hold:*

1. *If a set $A$ is dependent, then there is $B \subset A$ such that $B$ is superfluous in $AGT$.*
2. *A set $S$ is dependent iff there is $B \subset A$ such that $B$ is superfluous in $A$.*

Next, we concern joint knowledge and common knowledge. Knowledge relative to indiscernibility $ind(A \cup B)$ can be considered to be a joint knowledge of $A$ and $B$. $ind(A \cup B)$ is not greater than indiscernibility relations $ind(A)$ and $ind(B)$.

**Proposition 4.26** $K(A)X, K(B)X \subseteq K(A \cup B)X$.

Hence, a joint knowledge of a group of agents is not less than knowledge of any member of the group.

Knowledge relative to indiscernibility $ind(A \cap B)$ can be considered to be a common knowledge of $A$ and $B$. To discuss common knowledge, we have to admit non-transitive indiscernibility relations. Here, we introduce the following notation:

$ind(A) \circ ind(B) = ind(AB)$, where $\circ$ is the composition of relations,
$(ind(A) \cup ind(B))^* = ind(\{A, B\}^*)$.

Here, $\{A, B\}^*$ is the set of all finite sequences with elements from the set $\{A, B\}$.

For $S \in \{A, B\}^*$, $ind(S)$ is the composition of $ind(A)$ for all the elements $A$ of sequence $S$.

**Proposition 4.27** *The following hold:*

1. $ind(A), ind(B) \subseteq ind(AB)$
2. $ind(\{A, B\}^*) \subseteq ind(A \cap B)$
3. *If $ind(AB) = ind(BA)$, then $ind(\{A, B\}^*) = ind(AB)$*
4. *If $ind(A) \subseteq ind(B)$, then $ind(AC) \subseteq ind(BC)$ and $ind(CA) \subseteq ind(CB)$ for any $C \subseteq AGT$*
5. *If $AGT$ is independent, then $ind(\{A, B\}^*) = ind(A \cap B)$.*

Observe that for $S \in \{A, B\}^*$, the relation $ind(S)$ is not necessarily transitive.

**Proposition 4.28** $K(A \cap B)X \subseteq K(S)X$ *for any* $S \in \{A, B\}^*$.

Hence, common knowledge of $A$ and $B$ is included in the knowledge relative to composition of relations $ind(A)$ and $ind(B)$.

Orlowska defined a propositional language with a family of relative knowledge operators. Each operator is determined by a set of parameters interpreted as knowledge agents.

Let $CONAGT$ be a set of constants which are to be interpreted as sets of agents. We define the set $EXPAGT$ of agent expressions:

$CONAGT \subseteq EXPAGT$,
$A, B \in EXPAGT$ imples $-A, A \cup B, A \cap B \in EXPAGT$.

Let $VARPROP$ be a set of propositional variables. The set $FOR$ of formulas is the smallest set satisfying the following conditions:

$VARPROP \subseteq FOR$
$F, G \in FOR$ implies $\neg F, F \vee G, F \wedge G, F \rightarrow G, F \leftrightarrow G \in FOR$
$A \in EXPAGT, F \in FOR$ imply $K(A)F \in FOR$

The set $FOR$ is closed with respect to classical propositional connectives and knowledge operators determined by agent expressions.

Let an *epistemic system* $E = (U, AGT, \{ind(A)\}_{A \in AGT})$ be given. By a *model*, we mean a system $M = (E, m)$, where $m$ is a meaning function assigning sets of states to propositional variables, sets of agents to agent constants, and moreover $m$ satisfies the following conditions:

$m(p) \subseteq U$ for $p \in VARPROP$,
$m(A) \subseteq AGT$ for $A \in CONAGT$,
$m(A \cup B) = m(A) \cup m(B)$,
$m(A \cap B) = m(A) \cap m(B)$,
$m(-A) = -m(A)$.

We define inductively a family of set $ext_M\ F$ (extension of formula $F$ in model $M$) for any $F \in FOR$:

$ext_M\ p = m(p)$ for $p \in VARPROP$,
$ext_M\ (\neg F) = -ext_M\ F$,
$ext_M\ (F \vee G) = ext_M\ F \cup ext_M\ G$,
$ext_M\ (F \wedge G) = ext_M\ F \cap ext_M\ G$,
$ext_M\ (F \rightarrow G) = ext_M(\neg F \vee G)$,
$ext_M\ (F \leftrightarrow G) = ext_M((F \rightarrow G) \wedge (G \rightarrow F))$,
$ext_M\ K(A)F = K(m(A))ext_M\ F$.

We say that a formula $F$ is *true in a model* $M$ ($\models_M F$) iff $ext_M\ F = U$ and a formulas $F$ is *valid* ($\models F$) iff it is true in all models.

Observe that formulas of the form $K(A)F \rightarrow F$ and $(F \rightarrow G) \wedge K(A)F \rightarrow K(A)G$ are not valid. This means that if $F$ is known by an agent, then $F$ is not necessarily true, and agents do not know all the consequences of their knowledge. Thus, the system can avoid well-known paradoxes in epistemic logic.

In the following, we list some facts about knowledge of agents which can be expressed in the logic.

**Proposition 4.29** *The following hold:*

1. $\models (F \leftrightarrow G)$ *implies* $\models (K(A)F \leftrightarrow K(A)G)$
2. $\models F$ *implies* $\models K(A)F$
3. $\models K(A)F \rightarrow K(A)K(A \cup N)F$
4. $\models K(A \cup B)K(A)F \rightarrow K(A \cup B)F$
5. $\models (K(A)F \vee K(B)F) \rightarrow K(A \cup B)F$
6. $\models K(A \cap B)F \rightarrow K(A)F \wedge K(B)F$
7. $\models K(A)F \leftrightarrow K(A)(\neg F)$
8. $\models K(A)(K(A)F \rightarrow F)$

Here, (4) says that knowledge of a group of agents exceeds knowledge of a part of the group. (7) results from the fact that agents $A$ can tell extension of $F$ from its complement iff they can tell extension of $\neg F$ from its complement.

Observe that $\models F$ implies $\models K(A)F$. This fact is often considered to be a paradox of ideal knowers. However, it seems to be less paradoxical by using the interpretations of knowledge as the ability to decide the membership question.

It follows from the fact that the whole universe $U$ is $A$-definable for any $A$. In other words, whatever a perception ability of $A$ is (whatever $ind(A)$ is), equivalence classes of all the elements from $U$ over $U$.

Orlowska's logic is significantly different from standard epistemic logic (cf. Hintikka [28], Fagin et al. [29], Halpern et al. [30, 31]), although it is based on Kripke semantics. Her logic can in fact overcome several defects of epistemic logic, and can be considered to be an interesting alternative for a logic of knowledge.

Orlowska only developed the semantics for her logics, and complete axiomatizations are open. However, for applications to real problems, we have to investigate a proof theory for logics for reasoning about knowledge.

In Orlowska's approach, approximation operators are based on Pawlak's rough set theory, but they can be generalized in several ways. One of the interesting generalizations is to define knowledge operators relative with respect to arbitrary binary relations by using the generalized notion of approximation of a set; see Orlowska [25].

Let $R$ be a binary relation in a set $U$. For $x \in U$, we define a neighbourhood of $x$ with respect to $R$:

$$n_R(x) = \{y \in U \mid (x, y) \in R \text{ or } (y, x) \in R\}$$

Then, by lower (upper) approximation of a set $X \subseteq U$, we mean the union of those neighbourhoods which are included in $X$ (which have an element in common with $X$).

To define the respective knowledge operators, we assume that with every set $A$ of agents there is associated a relation $R(A)$ in the set $U$. The corresponding *epistemic structure* is $(U, AGT, \{R(A)\}_{A \subseteq AGT}, \{K(A)\}_{A \subseteq AGT})$, where $R : P(AGT) \rightarrow P(U \times U)$ assigns binary relations to sets of agents, and $K : P(AGT) \rightarrow P(U)$ is an operator such that $K(A)(X)$ is the union of the lower approximation of $X$ and the complement of the upper approximations of $X$ with respect to $R(A)$.

Thus, Kripke structures can be generalized with accessibility relations determined by sets of parameters as:

(KR) $\mathbf{K} = (W, PAR, RL, \{R(P)\}_{P \subseteq PAR, R \in REL})$

where $W$ is a non-empty set of worlds (or states, objects, etc.), $PAR$ is a non-empty set whose elements are called parameters, elements of set $REL$ are mapping $R : P(PAR) \rightarrow P(W \times W)$ which assign binary relation in set $W$ to subseteq of set $PAR$. Moreover, we assume that $R(P)$ satisfies the following conditions:

$R(\emptyset) = W \times W$

$R(P \cup Q) = R(P) \cap R(Q)$

The first condition says that the empty set of parameters does not enable us to distinguish any worlds. The second condition says that if we have more parameters then the relation is smaller, less worlds will be glue together.

We here observe that the axiomatization of logics based on Kripke models with relative accessibility relations is an open problem.

## 4.8 Logics for Knowledge Representation

Although Orlowska's logic is concerned with reasoning about knowledge, we can find several (modal) logics for knowledge representation in the literature, some of which are closely related to rough set theory.

In this section, we review these logics. The starting point of such approaches is an *information system* introduced by Pawlak [32]. An information system is a collection of pieces of information which have the form: object, a list of properties of the object. Object is anything that can be used in a subject position of a natural language sentence.

Although object can be composed and structured in information systems, they are treated as indivisible wholes. Properties of objects are expressed by attributes and their values.

By formal terms, an information system is determined by specifying a non-empty set $OB$ of objects, a non-empty set $AT$ of attributes, a family $\{VAL_a\}_{a \in AT}$ of sets of values of attributes, and an information function $f$ which assigns properties to objects.

There are two types of information functions. *Deterministic* information function is a function of the form $f : OB \times AT \rightarrow VAL = \bigcup \{VAL_a \mid a \in AT\}$ which assigns a value of attribute to object. It is assumed that for any $x \in OB$ and $a \in AT$ we have $f(x, a) \in VAL_a$.

Functions of this type determines properties of objects in deterministic way, namely property is uniquely assigned to object.

The other type of information function is *non-deterministic* information function of the form $f : OB \times AT \rightarrow P(VAL)$, which assigns a subset of the set of values of attribute to object.

Non-deterministic information function reflects incompleteness of information about properties of objects. The function says what is the range of possible values of every attribute for an object, but the value itself is not known.

An information system is defined as a structure of the form:

$\mathbf{S} = (OB, AT, \{VAL_a\}_{a \in AT}, f)$

If $f$ is a deterministic information function, then system **S** is called *deterministic information system*, and if $f$ is non-deterministic information function, then **S** is called *non-deterministic information system*, respectively.

Information about properties of objects is a basic explicit information included in information systems. From that information, we can derive some other information which is usually expressed in terms of binary relations in the set of objects.

Let $A \subseteq AT$ be a set of attributes. By an *indiscernibility relation* determined by a set $A$, we mean the relation $ind(A) \subseteq OB \times OB$ defined as:

$(x, y) \in ind(A)$ iff $f(x, a) = f(y, a)$ for all $a \in AT$

For the empty set of attributes, we assume $ind(\emptyset) = OB \times OB$. Thus, two objects stand in relation $ind(A)$ whenever they cannot be distinguished one from the other by means of properties determined by the attribute from the set $A$.

The following proposition states the basic properties of indiscrniblity relations:

**Proposition 4.30** *The following properties hold for* $ind$:

1. $ind(A)$ *is reflexive, symmetric and transitive.*
2. $ind(A \cup B) = ind(A) \cap ind(B)$.
3. $(ind(A) \cup ind(B))^* \subseteq ind(A \cap B)$.
4. $A \subseteq B$ *imples* $ind(B) \subseteq ind(A)$.

Here, (1) says that indiscernibility relations are equivalence relations. Equivalence class of $ind(A)$ consists of those objects which are indistinguishable up to attributes from the set $A$.

(2) says that discrimination power of the union of sets of attributes is better than that of the parts of the union. Consequently, the algebra $(\{ind(A)\}_{A \subseteq AT}, \cap)$ is a lower semilattice with the zero element $ind(AT)$.

Indiscernibility of objects plays a crucial role in many applications in which definability of the set of objects is important in terms of properties of single objects. Indiscernibility relations can be derived both from deterministic and non-deterministic information systems.

In connection with non-deterministic information systems, several other relations have been discussed; see Orlowska and Pawlak [33].

Let **S** be a non-deterministic information system. We define a family of *similarity relations* for $A \subseteq AT$, denoted $sim(A)$:

$(x, y) \in sim(A)$ iff $f(x, a) \cap f(y, a) \neq \emptyset$ for all $a \in A$.

We can also consider weak similarity of objects:

$(x, y) \in wsim(A)$ iff there is $a \in A$ such that $f(x, a) \cap f(y, a) \neq \emptyset$.

For some applications, negative similarity might be interesting:

$(x, y) \in nsim(A)$ iff $-f(x, a) \cap -f(y, a) \neq \emptyset$ for all $a \in A$.

Information inclusion of objects ($in(A)$) and weak information inclusion ($win(A)$) are defined as follows:

$(x, y) \in in(A)$ iff $f(x, a) \subseteq f(y, a)$ for all $a \in A$
$(x, y) \in win(A)$ iff there is $a \in A$ such that $f(x, a) \subseteq f(y, a)$.

Observe that the given relations are not independent, they satisfy the following conditions:

**Proposition 4.31** *The following conditions hold:*

*1.* $(x, y) \in in(A)$ *and* $(x, z) \in sim(A)$ *imply* $(y, z) \in sim(A)$.
*2.* $(x, y) \in ind(A)$ *implies* $(x, y) \in in(A)$.
*3.* $(x, y) \in in(A)$ *and* $(y, x) \in in(A)$ *imply* $(x, y) \in ind(A)$.

It is natural to connect information systems and modal logics. A Kripke structure is of the form $\mathbf{K} = (W, R)$, where $W$ is a non-empty set of possible worlds or states, and $R \subseteq W \times W$ is a binary relation called a accessibility relation.

With every information system of the form (**S**), we can associate a corresponding Kripke structure $\mathbf{K}(\mathbf{S})$, where set $OB$ is considered to be the universe and relations determined by the system is considered to be accessibility relations (cf. [34]).

Accessibility relations for indiscernibility are assumed to be equivalence relations. Similarity relations are reflexive and symmetric, and information inclusion is reflexive and transitive.

However, we have to assume that a family of indiscernibility relations is closed under intersection, that conditions (1), (2), (3) in Proposition 4.31 are satisfied in $\mathbf{K}(\mathbf{S})$, providing relationships between indiscernibility, similarity and information inclusion.

Several modal logics for reasoning about objects in information systems have been proposed. Fariñas del Cerro and Orlowska [35] developed data analysis logic $DAL$. The logic can handle inferences in the presence of incomplete information.

Data analysis can be understood as a process of obtaining patterns in a set of data items. They considered two main tasks involved in data analysis in $DAL$, namely

(1) to aggregate data into sets according to their properties,
(2) to define properties adequate for characterization of sets of data.

Obviously, these two tasks are necessary for data analysis.

$DAL$ defined formal counterparts of sets of data and properties. Namely, sets of data are defined by means of the language of $DAL$ and properties are defined by means of relational expressions.

Thus, data are identified with a non-empty set of objects and a family of equivalence relations on this set. Objects will be interpreted as data items and relations correspond to properties of data items. Each property induces an equivalence relation such that an equivalence class of the relation consists of those objects which are the same with respect to this property.

The language of $DAL$ includes modal operators interpreted as operations of lower and upper approximations determined by indiscernibility relations. Semantics of $DAL$ is given by Kripke structures with indiscernibility relations. An algebraic structure is assumed in the set of indiscernibility relations, namely the set is closed under intersection and transitive closure of union of relations.

Expressions of $DAL$ are built with symbols from the following pairwise disjoint sets:

$VARPROP$: a denumerable set of propositional variables
$\{IND_i\}$: a denumerable set of relational constants ($i$ is a natural number)
$\{\cap, \cup^*\}$: the set of relational operations of intersection and transitive closure of union
$\{\neg, \vee, \wedge, \rightarrow\}$: the set of classical propositional operations of negation, disjunction and implication
$\{[\ ], \langle\ \rangle\}$: the set of modal operators.

The set $EREL$ of relational expressions is the smallest set satisfying the following conditions:
$CONREL \subseteq EREL$
$IND_1, IND_2 \in EREL$ implies $IND_1 \cap IND_2, IND_1 \cup^* IND_2 \in EREL$

The set $FOR$ of formulas is the smallest set such that:
$VARPROP \subseteq FOR$
$F, G \in FOR$ implies $\neg F, F \vee G, F \wedge G, F \rightarrow G \in FOR$
$F \in FOR, IND \in EREL$ imply $[IND]F, \langle IND \rangle F \in FOR$.

A semantics for $DAL$ is given by the Kripke model of the form:

(MD) $M = (OB, \{S_p\}_{p \in VARPROP}, \{ind_i\}, m)$

where $OB$ is a non-empty set of objects, for any propositional variable $p$ set $S_p$ is a subset of $OB$, for any natural number $i$ $ind_i$ is an equivalence relation in$OB$, and $m$ is a meaning function satisfying the following conditions:

(m1) $m(p) = S_p$ for $p \in VARPOP, m(IND_i) = ind_i$ for any $i$
(m2) $m(IND_1 \cap IND_2) = m(IND_1) \cap m(IND_2)$
(m3) $m(IND_1 \cup^* IND_2) = m(IND_1) \cup^* m(IND_2)$
(m4) $m(\neg F) = -m(F)$
(m5) $m(F \vee G) = m(F) \vee m(G)$
(m6) $m(F \wedge G) = m(F) \wedge m(G)$
(m7) $m(F \rightarrow G) = m(F) \rightarrow m(G)$

(m8) $m([IND]F) = \{x \in OB \mid$ for all $y \in OB$ if $(x, y) \in m(IND)$ then $y \in m(F))$

(m9) $m(\langle IND\rangle F) = \{x \in OB \mid$ there is $y \in OB$ such that $(x, y) \in m(IND)$ and $y \in m(F))$

A formula is true in model $M$ ($\models_M F$) iff $m(F) = OB$. A formula $F$ is valid ($\models F$) iff $F$ is true in all models.

In the semantics for $DAL$ above, formulas are interpreted as subsets of the set of objects from a model, classical propositional operations are the counterparts of set-theoretical operations, and the modal operators $[IND]$ and $\langle IND\rangle$ correspond to operations of lower and upper approximation, respectively, with respect to indiscernibility relation $m(IND)$.

**Proposition 4.32** *The following hold:*

(1) $\models_M F \rightarrow [IND]F$ *iff* $m(F)$ *is* $m(IND)$*-definable*
(2) $\models_M [IND]F$ *iff* $POS(m(IND))m(F) = OB$
(3) $\models_M \neg[IND]F$ *iff* $POS(m(IND))m(F) = \emptyset$
(4) $\models_M \neg\langle IND\rangle F$ *iff* $NEG(m(IND))m(F) = OB$
(5) $\models_M \langle IND\rangle F$ *iff* $NEG(m(IND))m(F) = \emptyset$
(6) $\models_M \langle IND\rangle F \wedge \langle IND\rangle\neg F$ *iff* $BOR(m(IND))m(F) = OB$
(7) $\models_M \langle IND\rangle F \wedge [IND]\neg F$ *iff* $BOR(m(IND))m(F) = \emptyset$

Since indiscernibility relations, their intersections and transitive closures of union are equivalence relations, operators $[IND]$ and $\langle IND\rangle$ are S5 operators of necessity and possibility, respectively. Hence, all the formulas which are substitutions of theorems of S5 are valid in $DAL$.

**Proposition 4.33** *The following hold.*

1. $\models [IND_1]F \vee [IND_2]F \rightarrow [IND_1 \cap IND_2]F$
2. $\models [IND_1 \cup^* IND_2]F \rightarrow [IND_1]F \wedge [IND_2]F$

Note here that it is an open problem of finding a Hilbert style axiomatization of $DAL$.

Model of the form (MD) presented above is said to be a model with *local agreement* of indiscernibility relation whenever for any relations $ind_1$ and $ind_2$, and for any object $x$, the equivalence class of $ind_1$ determined by $x$ is included in the equivalence class of $ind_2$ determined by $x$, or conversely, the class of $ind_2$ is included in the respective class of $ind_1$.

The complete axiomatization of $DAL$ with respect to the class of models with local agreement is the following:

(D1) All formulas having the form of classical propositional tautology
(D2) $[IND](F \to G) \to ([IND]F \to [IND]G)$
(D3) $[IND]F \to F$
(D4) $\langle IND \rangle F \to [IND]\langle IND \rangle F$
(D5) $[IND \cup^* IND_2]F \to [IND_1]F \wedge [IND_2]F$
(D6) $(([IND_1]F \to [IND_2]F) \wedge ([IND_1]F \to [IND_2]F)) \to ([IND_2]F \to [IND_2 \cup^* IND_3]F)$
(D7) $[IND_1]F \vee [IND_2]F \to (IND_1 \cap IND_2]F$
(D8) $(([IND_1]F \to [IND_3]F) \wedge ([IND_2]F \to [IND_3]F)) \to ([IND_1 \cap IND_2]F \to [IND_3]F)$.

The rules of inference ae *modus pones* and necessitation for all operators $[IND]$, i.e., $A/[IND]A$. Axioms (D1)–(D4) are the standard axioms of modal logic S5.

Fariñas del Cerro and Orlowska [35] proved completeness of $DAL$. $DAL$ can be extended in various ways. Such extensions involve other operations on relations. For example, we can add the composition of relations. Another possibility is to assume relations which are not necessarily equivalence relations.

Balbiani [36] solved the axiomatization problem of $DAL$ and proposed $DAL^{\cup}$ with a completeness result. He used a Kripke semantics with $\cup$-relative accessibility relation for $DAL^{\cup}$. See Balibiani [36] for details.

To reason about non-deterministic information logic $NIL$ (non-deterministic information logic) has been introduced with modal operators determined by similarity and information inclusion of objects; see Orlowska and Pawlak [33].

The set of formulas of the language of logic $NIL$ is the least set including a set $VARPOP$ of propositional variables and closed under classical propositional operations and modal operations $[SIM]$, $\langle SIM \rangle$, $[IN]$, $\langle IN \rangle$, $[IN^{-1}]$, $\langle IN^{-1} \rangle$, where $SIM$ and $IN$ are relational constants interpreted as similarity and informational inclusion and $IN^{-1}$ denotes the converse of relation $IN$.

Kripke models of logic $NIL$ are systems of the form:
(MN) $M = (OB, \{S_p\}_{p \in VARPOP}, \{sim, in\}, m)$

where $OB$ is a non-empty set of objects, $S_p \subseteq OB$ for every $p \in VARPROP$, $sim$ is a reflexive and transitive relation in $OB$, and moreover, we assume that the relations satisfy the condition:

(n1) If $(x, y) \in in$ and $(x, z) \in sim$, then $(y, z) \in sim$.

Meaning function $m$ is defined as follows: $m(p) = S_p$ for $p \in VARPROP$, $m(SIM) = sim, m(IN) = in, m(IN^{-1}) = m(IN)^{-1}$; for complex formulas built with classical propositional operations $m$ is given by conditions (m4),..., (m9), where $IND$ is replaced by $SIM, IN, IN^{-1}$, respectively.

Set of axioms of logic $NIL$ consists of axioms of logic B for formulas with operators $[SIM]$ and $\langle SIM \rangle$, axioms of logic S4 for formulas with operators $[IN]$, $\langle IN \rangle$, $[IN^{-1}]$, $\langle IN^{-1} \rangle$, and the axiom corresponding to condition (n1):

(N1) $[SIM]F \to [IN^{-1}][SIM][IN]F$

Rules of inference are *modus ponens* and necessitation for the three modal operators of the language.

Vakarelov [37] proved that models of $NIL$ represent adequately non-deterministic information systems, as the following proposition:

**Proposition 4.34** *For any model of the form (MN) there is a non-deterministic information system of the form* (**S**) *with the same set* $OB$ *of objects such that for any* $x, y \in OB$ *we have:*

$(x, y) \in sim$ *iff* $f(x, a) \cap f(y, a) \neq \emptyset$ *for all* $a \in AT$
$(x, y) \in sim$ *iff* $f(x, a) \subseteq f(y, a)$ *for all* $a \in AT$

Vakarelov [38] introduced information logic $IL$ as an extension of $NIL$. The language of $IL$ includes the operators of $NIL$ and $[IND]$ and $\langle IND \rangle$ determined by indiscernibility relation.

In the corresponding Kripke model, it is assumed that indiscernibility relation $ind$ is an equivalence relation and relations $sim, in, ind$ satisfy condition (n1) and the following conditions:

(n2) $ind \subseteq in$
(n3) $in \cap in^{-1} \subseteq ind$

Axioms of $IL$ include axioms of $NIL$, axioms of S5 for $[IND]$ and $\langle IND \rangle$, and the following axiom corresponding to (n2):

$[IN]F \rightarrow [IND]F$.

Note that condition (n3) is not expressible by means of a formula of $IL$.

Vakarelov proved completeness of the given set of axioms for the class of models satisfying conditions (n1) and (n2) and the class of models satisfying conditions (n1), (n2) and (n3). He also introduced some other information logics which correspond both to deterministic and non-deterministic information systems.

Logic $NIL$ can reason about objects which are incompletely defined. Here, we understand incompleteness as lack of definite information about values of attributes for objects. By using modal operators, we can compare objects with respect to informational inclusion and similarity.

In logic $IL$, it is possible to deal with two kinds of incomplete information: objects are defined up to indiscernibility, and their properties are specified nondeterministically.

Vakarelov also investigated a duality between Pawlak's knowledge representation system and certain information systmes of logical type, called *bi-consequence systems*. He developed a complete modal logic $INF$ for some informational relations; see Vakarelov [39].

Konikowska [40] proposed a modal logic for reasoning about relative similarity based on the idea of rough set theory. She presented a Kripke semantics and a Gentzen system and proved a completeness result.

As discussed in this section, logics for data analysis and knowledge representation are also of interest to describe incompleteness of knowledge based on the concept of indiscernibility. We can point out that they are more powerful than current knowledge representation languages used in AI.

However, on the theoretical side, the problem of axiomatization remains open for many such logics. The lack of practical proof methods is serious, and detailed studies are needed.

## References

1. Iwinski, T.: Algebraic approach to rough sets. Bull. Pol. Acad. Math. **37**, 673–683 (1987)
2. Pomykala, J., Pomykala, J.A.: The stone algebra of rough sets. Bull. Pol. Acad. Sci., Math. **36**, 495–508 (1988)
3. Yao, Y., Lin, T.: Generalization of rough sets using modal logics. Intell. Autom. Soft Comput. **2**, 103–120 (1996)
4. Liau, C.-J.: An overview of rough set semantics for modal and quantifier logics. Int. J. Uncertain. Fuzziness Knowl. -Based Syst. **8**, 93–118 (2000)
5. de Caro, F.: Graded modalities II. Stud. Logica **47**, 1–10 (1988)
6. Fattorosi-Barnaba, M., de Caro, F.: Graded modalities I. Stud. Logica **44**, 197–221 (1985)
7. Fattorosi-Barnaba, M., de Caro, F.: Graded modalities III. Stud. Logica **47**, 99–110 (1988)
8. Murai, T., Miyakoshi, M., Shinmbo, M.: Measure-based semantics for modal logic. In: Lowen, R., Roubens, M. (eds.) Fuzzy Logic: State of the Arts, pp. 395–405. Kluwer, Dordrecht (1993)
9. Murai,T., Miyakoshi, M., Shimbo, M.: Soundness and completeness theorems between the Dempster-Shafer theory and logic of belief. In: Proceedings of 3rd FUZZ-IEEE (WCCI), pp. 855–858 (1994)
10. Fattorosi-Barnaba, M., Amati, G.: Modal operators with probabilistic interpretations I. Stud. Logica **46**, 383–393 (1987)
11. Wong, S., Ziarko, W.: Comparison of the probabilistic approximate classification and the fuzzy set model. Fuzzy Sets Syst. **21**, 357–362 (1987)
12. Pagliani, P.: Rough sets and Nelson algebras. Fundam. Math. **27**, 205–219 (1996)
13. Järvinen, J., Pagliani, P., Radeleczki, S.: Information completeness in Nelson algebras of rough sets induced by quasiorders. Stud. Logica **101**, 1073–1092 (2013)
14. Rasiowa, H.: An Algebraic Approach to Non-Classical Logics. North-Holland, Amsterdam (1974)
15. Vakarelov, D.: Notes on constructive logic with strong negation. Stud. Logica **36**, 110–125 (1977)
16. Iturrioz, L.: Rough sets and three-valued structures. In: Orlowska, E. (ed.) Logic at Work: essays Dedicated to the Memory of Helena Rasiowa, pp. 596–603. Physica-Verlag, Heidelberg (1999)
17. Sendlewski, A.: Nelson algebras through Heyting ones I. Stud. Logica **49**, 105–126 (1990)
18. Akama, S., Murai, T.: Rough set semantics for three-valued logics. In: Nakamatsu, K., Abe, J.M. (eds.) Advances in Logic Based Intelligent Systems, pp. 242–247. IOS Press, Amsterdam (2005)
19. Avron, A., Konikowska, B.: Rough sets and 3-valued logics. Stud. Logica **90**, 69–92 (2008)
20. Avron, A., Lev, I.: Non-deterministic multiple-valued structures. J. Logic Comput. **15**, 241–261 (2005)
21. Gentzen, G.: Collected papers of Gerhard Gentzen. In: Szabo, M.E. (ed.) North-Holland, Amsterdam (1969)
22. Düntsch, I.: A logic for rough sets. Theoret. Comput. Sci. **179**, 427–436 (1997)
23. Akama, S., Murai, T., Kudo, Y.: Heyting-Brouwer rough set logic. In: Proceedings of KSE2013, pp. 135–145. Hanoi, Springer, Heidelberg (2013)

24. Akama, S., Murai, T., Kudo, Y.: Da Costa logics and vagueness. In: Proceedings of GrC2014, Noboribetsu, Japan (2014)
25. Orlowska, E.: Kripke models with relative accessibility relations and their applications to inferences from incomplete information. In: Mirkowska, G., Rasiowa, H. (eds.) Mathematical Problems in Computation Theory, pp. 327–337. Polish Scientific Publishers, Warsaw (1987)
26. Orlowska, E.: Logical aspects of learning concepts. Int. J. Approximate Reasoning **2**, 349–364 (1988)
27. Orlowska, E.: Logic for reasoning about knowledge. Zeitschrift für mathematische Logik und Grundlagen der Mathematik **35**, 559–572 (1989)
28. Hintikka, S.: Knowledge and Belief. Cornell University Press, Ithaca (1962)
29. Fagin, R., Halpern, J., Moses, Y., Vardi, M.: Reasoning about Knowledge. MIT Press, Cambridge. Mass (1995)
30. Halpern, J., Moses, Y.: Towards a theory of knowledge and ignorance: preliminary report. In: Apt, K. (ed.) Logics and Models of Concurrent Systems, pp. 459–476. Springer, Berlin (1985)
31. Halpern, J., Moses, Y.: A theory of knowledge and ignorance for many agents. J. Logic Comput. **7**, 79–108 (1997)
32. Pawlak, P.: Information systems: theoretical foundations. Inf. Syst. **6**, 205–218 (1981)
33. Orlowska, E., Pawlak, Z.: Representation of nondeterministic information. Theoret. Comput. Sci. **29**, 27–39 (1984)
34. Orlowska, E.: Kripke semantics for knowledge representation logics. Stud. Logica **49**, 255–272 (1990)
35. Fariñas del Cerro, L., Orlowska, E.: DAL-a logic for data analysis. Theoret. Comput. Sci. **36**, 251–264 (1985)
36. Balbiani, P.: A modal logic for data analysis. In: Proceedings of MFCS'96, pp. 167–179. LNCS 1113, Springer, Berlin
37. Vakarelov, D.: Abstract characterization of some knowledge representation systems and the logic $NIL$ of nondeterministic information. In: Skordev, D. (ed.) Mathematical Logic and Applications. Plenum Press, New York (1987)
38. Vakarelov, D.: Modal logics for knowledge representation systems. Theoret. Comput. Sci. **90**, 433–456 (1991)
39. Vakarelov, D.: A modal logic for similarity relations in Pawlak knowledge representation systems. Stud. Logica **55**, 205–228 (1995)
40. Konikowska, B.: A logic for reasoning about relative similarity. Stud. Logica **58**, 185–228 (1997)

# Chapter 5
# A Granularity-Based Framework of Reasoning

**Abstract** This chapter presents a granularity-based framework of deduction, induction, and abduction using variable precision rough set models proposed by Ziarko and measure-based semantics for modal logic proposed by Murai et al. This is of special importance as a general approach to reasoning based on rough set theory. We also discuss non-monotonic reasoning, association rules in conditional logic, and background knowledge.

## 5.1 Deduction, Induction and Abduction

Reasoning processes in our daily life consist of various styles of reasoning under uncertainty, such as logical reasoning with some non-monotonicity, probabilistic reasoning, and reasoning with ambiguity and vagueness; for example, implying conclusions logically from information we currently possess, finding rules from observations, and speculate reasons behind observed (or reported) facts.

In general, logical aspects of these types of reasoning processes are divided into the following three categories:

- **Deduction**: A reasoning process for concluding specific facts from general rules.
- **Induction**: A reasoning process for providing general rules from specific facts.
- **Abduction**: A reasoning process for providing hypotheses that explain the given facts.

Moreover, when we consider these types of reasoning processes, we consider not all possible scenarios or situations that match the propositions used in them, but some typical scenarios or situations.

For example, suppose we consider the following deduction: from the propositions "the sun rises in the east" and "if the sun rises in the east, then the sun sets in the west," we conclude that "the sun sets in the west".

In this deduction process, we do not consider all days when the sun rose in the east, and we may consider only a small number of examples of days when the sun rose in the east as typical situations. Moreover, because the sun set in the west on any typical day when the sun rose in the east, we conclude that the sun sets in the west.



S. Akama et al., *Reasoning with Rough Sets*, Intelligent Systems Reference Library 142, https://doi.org/10.1007/978-3-319-72691-5_5

In other words, typical situations in which the sun rises in the east are also typical situations in which the sun sets in the west. This example indicates that considering the relationship between typical situations captures aspects of deduction, induction, and abduction in our daily life.

In this chapter, we consider the semantic characterization of deduction, induction, and abduction by the possible world semantics of modal logic. In possible world semantics, each non-modal sentence that represents a fact is characterized by its truth set, i.e., the set of all possible worlds in which the non-modal sentence is true in the given model.

We consider the truth set of a non-modal sentence as the correct representation of the given fact. However, as we have discussed, we need to treat typical situations related to facts, and treating only the truth sets of nonmodal sentences that represent facts is not suitable, because these truth sets correspond to all situations that match the facts.

Thus, we must represent typical situations based on some theory. To represent typical situations about the facts, we consider introducing rough set theory to the possible world semantics of modal logic as reviewed in Chap. 4.

Combining the above discussions, we propose a unified framework of deduction, induction, and abduction using granularity based on VPRS models and measure-based semantics for modal logic. As reviewed in Chap. 2, VPRS is based on the majority relation.

Let $U$ be a universe and $X, Y$ be any subsets of $U$. The majority inclusion relation is defined by the measure $c(X, Y)$ of the relative degree of misclassification of $X$ with respect to $Y$.

Formally, the majority inclusion relation $\overset{\beta}{\subseteq}$ with a fixed precision $\beta \in [0, 0.5)$ is defined using the relative degree of misclassification as follows:

$X \overset{\beta}{\subseteq} Y$ iff $c(X, Y) \leq \beta$

where the precision $\beta$ provides the limit of permissible misclassification.

Let $X \subseteq U$ be any set of objects, $R$ be an indiscernibility relation on $U$, and the degree $\beta \in [0, 0.5)$ be a precision. The $\beta$-lower approximation $\underline{R}_\beta(X)$ and the $\beta$-upper approximation $\overline{R}_\beta(X)$ of $X$ are defined as follows:

$$\underline{R}_\beta(X) = \{x \in U \mid [x]_R \overset{\beta}{\subseteq} X\} = \{x \in U : c([x]_R, X) \leq \beta\},$$
$$\overline{R}_\beta(X) = \{x \in U \mid c([x]_R, X) < 1 - \beta\}.$$

As mentioned in Chap. 2, the precision $\beta$ represents the threshold degree of misclassification of elements in the equivalence class $[x]_R$ to the set $X$. Thus, in VPRS, misclassification of elements is allowed if the ratio of misclassification is less than $\beta$. Note that the $\beta$-lower and -upper approximations with $\beta = 0$ correspond to Pawlak's lower and upper approximations.

**Table 5.1** Some properties of $\beta$-lower and upper approximations

| Properties | Conditions | $\beta = 0$ | $0 < \beta < 0.5$ |
|---|---|---|---|
| Df$\Diamond$ | $\overline{R}_\beta(X) = \underline{R}_\beta(X^c)^c$ | o | o |
| M | $\underline{R}_\beta(X \cap Y) \subseteq \underline{R}_\beta(X) \cap \underline{R}_\beta(Y)$ | o | o |
| C | $\underline{R}_\beta(X) \cap \underline{R}_\beta(Y) \subseteq \underline{R}_\beta(X \cap Y)$ | o | × |
| N | $\underline{R}_\beta(U) = U$ | o | o |
| K | $\underline{R}_\beta(X^c \cup Y) \subseteq (\underline{R}_\beta(X)^c \cup \underline{R}_\beta(Y))$ | o | × |
| D | $\underline{R}_\beta(X) \subseteq \overline{R}_\beta(X)$ | o | o |
| P | $\underline{R}_\beta(\emptyset) = \emptyset$ | o | o |
| T | $\underline{R}_\beta(X) \subseteq X$ | o | × |
| B | $X \subseteq \underline{R}_\beta(\overline{R}_\beta(X))$ | o | o |
| 4 | $\underline{R}_\beta(X) \subseteq \underline{R}_\beta(\underline{R}_\beta(X))$ | o | o |
| 5 | $\overline{R}_\beta(X) \subseteq \underline{R}_\beta(\overline{R}_\beta(X))$ | o | o |

Table 5.1 represents some properties of the $\beta$-lower and -upper approximations. The symbols "o" and "×" indicate whether a property is satisfied (o) or may not be satisfied (×) in the case of $\beta = 0$ and $0 < \beta < 0.5$, respectively.

For example, by the definition of the $\beta$-lower approximation, it is easy to confirm that the property T. $\underline{R}_\beta(X) \subseteq X$ is not guaranteed to be satisfied in the case of $0 < \beta < 0.5$. Note that symbols assigned to properties such that T correspond to axiom schemas in modal logic, as will be mentioned later.

## 5.2 Measure-Based Semantics

One of the interesting extensions of modal logic approach to rough sets is what we call *measure-based semantics*. It was developed by Murai et al. in [1, 2]. Instead of using accessibility relations to interpret modal sentences, measure-based semantics of modal logic uses fuzzy measures.

Let $L_{ML}(\mathscr{P})$ be the language of modal logic $ML$ constructed from a infinite set of atomic sentences $\mathscr{P} = \{p_1, p_2, ...\}$. We say that a sentence is modal if it contains at least one modal operator; else it is non-modal.

A function $\mu : 2^U \to [0, 1]$ is called a *fuzzy measure* on $U$ if the function $\mu$ satisfies the following three conditions:

(1) $\mu(U) = 1$,
(2) $\mu(\emptyset) = 0$,
(3) $\forall XY \subseteq U(X \subseteq Y \Rightarrow \mu(X) \leq \mu(Y))$,

where $2^U$ represents the power set of $U$.

Formally, a fuzzy measure model $M_\mu$ is the following triple,
$M_\mu = \langle U, \{\mu_x\}_{x \in U}, v \rangle$

where $U$ is a set of possible worlds, and $v$ is a valuation. $\{\mu_x\}_{x \in U}$ is a class of fuzzy measures $\mu_x$ assigned to all possible worlds $x \in U$.

In measure-based semantics of modal logic, each degree $\alpha \in [0, 1]$ of fuzzy measures corresponds to a modal operator $\Box_\alpha$. Thus, fuzzy measure models can provide semantics of multi-modal logic with modal operators $\Box_\alpha$ $(\alpha \in [0, 1])$. However, we here fix a degree $\alpha$ and consider $\alpha$-level fuzzy measure models that provide semantics of modal logic with the two modal operators $\Box$ and $\Diamond$.

Similar to the case of Kripke models, $M_\mu, x \models p$ indicates that the sentence $p$ is true at the possible world $x \in U$ by the $\alpha$-level fuzzy measure model $M_\mu$. Interpretation of non-modal sentences is identical to that in Kripke models. On the other hand, to define the truth value of modal sentences at each world $x \in U$ in the $\alpha$-level fuzzy measure model $M_\mu$, we use the fuzzy measure $\mu_x$ assigned to the world $x$ instead of accessibility relations.

Interpretation of modal sentences $\Box p$ at a world $x$ is defined as follows:

$$M_\mu, x \models \Box p \Leftrightarrow \mu_x(\|p\|^{M_\mu}) \geq \alpha$$

where $\mu_x$ is the fuzzy measure assigned to $x$. By this definition, interpretation of modal sentences $\Diamond p$ is obtained by dual fuzzy measures as follows:

$$M_\mu, x \models \Diamond p \Leftrightarrow \mu_x^*(\|p\|^{M_\mu}) > 1 - \alpha$$

where the dual fuzzy measure $\mu_x^*$ of the assigned fuzzy measure $\mu_x$ is defined as $\mu_x^* = 1 - \mu_x(X^c)$ for any $X \subseteq U$.

Note that the modal systems EMNP is sound and complete with respect to the class of all $\alpha$-level fuzzy measure models [1, 2]. EMNP consists of all inference rules and axiom schemas of propositional logic and the following inference rules and axiom schemas:

(Df$\Diamond$) $\Diamond p \leftrightarrow \neg\Box\neg p$
(RE) $p \leftrightarrow q / \Box p \leftrightarrow \Box q$
(M) $\Box(p \wedge q) \rightarrow (\Box p \wedge \Box q)$
(N) $\Box\top$
(P) $\neg\Box\bot$

Next, we introduce $\alpha$-level fuzzy measure models based on background knowledge to characterize typical situations as a modality of modal logic using granularity based on VPRS and measure-based semantics for modal logic.

As a basis of reasoning using granularity based on VPRS and measure-based semantics, suppose that we have a Kripke model $M = \langle U, R, v \rangle$ consisting of the given approximation space $(U, R)$ and a valuation $v$. In the Kripke model $M$, any non-modal sentence $p$ that represents a fact is characterized by its truth set $\|p\|^M$.

When we consider the fact represented by the non-modal sentence $p$, we may not consider all possible worlds in the truth set $\|p\|^M$. In such cases, we often consider only typical situations about the fact $p$.

To capture such typical situations, we examine the lower approximation of the truth set $\|p\|$ by the indiscernibility relation $R$, and consider each possible world in the lower approximation of the truth set $\|p\|$ as a typical situation about $p$ based on background knowledge about $U$.

Moreover, it may be useful to consider situations that are not typical about the facts as exceptions to typical situations. Here, we represent this characteristic using $\beta$-lower approximations of the truth sets of sentences that represent facts.

Thus, using background knowledge from the Kripke model $M$, we can consider the following two sets of possible worlds about a fact $p$:

- $\|p\|^M$: correct representation of fact $p$
- $\underline{R}_\beta(\|p\|^M)$: the set of typical situations about $p$ (situations that are not typical may also be included)

Using the given Kripke model as background knowledge, we define an $\alpha$-level fuzzy measure model to treat typical situations about facts as $\beta$-lower approximations in the framework of modal logic.

**Definition 5.1** Let $M = \langle U, R, v\rangle$ be a Kripke model that consists of an approximation space $(U, R)$ and a valuation function $v : P \times U \to \{0, 1\}$ and $\alpha \in (0.5, 1]$ be a fixed degree. An $\alpha$-level fuzzy measure model $M_\alpha^R$ based on background knowledge is the following triple:

$$M_\alpha^R = \langle U, \{\mu_x^R\}_{x\in U}, v\rangle$$

where $U$ and $v$ are the same as in $M$. The fuzzy measure $\mu_x^R : 2^U \to [0, 1]$ assigned to each $x \in U$ is a probability measure based on the equivalence class $[x]_R$ with respect to $R$, defined by

$$\mu_x^R(X) = \frac{|[x]_R \cap X|}{|[x]_R|}, \ \forall X \subseteq U.$$

Similar to the case of Kripke models, we denote that a sentence $p$ is true at a world $x \in U$ by an $\alpha$-level fuzzy measure model $M_\alpha^R$ by $M_\alpha^R, x \models p$. Truth valuation of modal sentences is defined as

$$M_\alpha^R, x \models \Box p \Leftrightarrow \mu_x^R(\|p\|^{M_\alpha^R}) \geq \alpha,$$

$$M_\alpha^R, x \models \Diamond p \Leftrightarrow \mu_x^R(\|p\|^{M_\alpha^R}) > 1 - \alpha.$$

We also denote the truth set of a sentence in the $\alpha$-level fuzzy measure model $M_\alpha^R$ by $\|p\|$, which is defined by

$$\|p\| = \{x \in U : M_\alpha^R, x \models p\}.$$

The constructed $\alpha$-level fuzzy measure model $M_\alpha^R$ from the given Kripke model $M$ has the following good properties:

**Theorem 5.1** *Let $M$ be a finite Kripke model such that it accessibility relation $R$ is an equivalence relation and $M_\alpha^R$ be the $\alpha$-level fuzzy measure model based on the background knowledge $M$. For any non-modal sentence $p \in L_{ML}(\mathscr{P})$ and any sentence $q \in L_{ML}(\mathscr{P})$, the following equations are satisfied:*

*(1)* $\|p\|^{M_\alpha^R} = \|p\|^M$
*(2)* $\|\Box q\|^{M_\alpha^R} = \underline{R}_{1-\alpha}(\|q\|^{M_\alpha^R})$
*(3)* $\|\Diamond q\|^{M_\alpha^R} = \overline{R}_{1-\alpha}(\|q\|^{M_\alpha^R})$

*Proof* (1) is clear from the definition of $\models$.

For (2), it is enough to show that any sentence $q \in L_{ML}(\mathscr{P})$ $M_\alpha^R$ $x \models \Box q$ holds if and only if $x \in \underline{R}_{1-\alpha}(\|q\|^M)$. Suppose $M_\alpha^R, x \models \Box q$ holds. By Definition 5.1, we have $\mu_x^R(\|r\|^{M_\alpha^R}) \geq \alpha$. By the definition of the relative degree of misclassification $c(X, Y)$ in $VPRS$ and the definition of the fuzzy measure $\mu_\alpha^R$, the property $\mu_x^R(\|q\|^M) \geq \alpha$ holds if and only if $c([x]_R, \|q\|^M) \leq 1 - \alpha$. Therefore, we have $x \in \underline{R}_{1-\alpha}(\|q\|)$. (3) is also similarly proved.

We here intend to use $\alpha$-level fuzzy measure models $M_\alpha^R$ as the basis of a unified framework for deduction, induction and abduction based on the concept of typical situations of facts and rules used in these reasoning processes.

Thus, as we discussed in Sect. 5.1, we represent facts and rules in reasoning processes as non-modal sentences and typical situations of facts and rules as lower approximations of truth sets of non-modal sentences.

From (1) and (2) in Theorem 5.1, the $\alpha$-level measure model $M_\alpha^R$ based on background knowledge $M$ exhibits the characteristics of correct representations of fact by the truth sets of non-modal sentences and typical situations of the facts by the $(1-\alpha)$-lower approximations of truth sets of non-modal sentences.

Thus, we can read a modal sentence $\Box p$ as "typically $p$" and represent the relationship between typical situations by modal sentences. This indicates that the models $M_\alpha^R$ are a sufficient basis for a unified framework for deduction, induction, and abduction.

Moreover, we have the following soundness properties of systems of modal logic with respect to the class of all $\alpha$-level fuzzy measure models based on background knowledge.

**Theorem 5.2** *For any $\alpha$-level fuzzy measure model $M_\alpha^R$ based on any finite Kripke model $M$ such that its accessibility relation $R$, the following properties are satisfied in the case of $\alpha = 1$ and $\alpha \in (0.5, 1)$, respectively:*

*If $\alpha = 1$, then all theorems of system S5 are true in $M_\alpha^R$,*
*If $\alpha \in (0.5, 1)$, then all theorems of EMND45 are true in $M_\alpha^R$,*

*where system EMND5 consists of the inference rules and axioms of the system EMNP and the following axiom schemas: D: $\Box p \to \Diamond p$, $4 : \Box p \to \Box\Box p$, $5 : \Diamond p \to \Box\Diamond p$.*

*Proof* It is clear from the correspondence relationship between axiom schemas and the properties of $\beta$-lower approximations shown in Table 5.1.

Theorem 5.2 indicates that the properties of $\alpha$-level fuzzy measure models based on background knowledge depend on the degree of $\alpha$. If we fix $\alpha = 1$, we do not allow any exception in typical situations; else, we allow some exceptions depending on $\alpha$.

This is because, if $\alpha = 1$, any $\alpha$-level fuzzy measure models $M_\alpha^R$ based on background knowledge satisfy the axiom schema **T**.$\Box p \to p$; else, $M_\alpha^R$ does not satisfy **T**. Thus, if $\alpha \in (0.5, 1)$, a non-modal proposition $p$ and a possible world $x \in U$ may exist such that $x \in \|\Box p\|^{M_\alpha^R}$ but $x \notin \|p\|^{M_\alpha^R}$; i.e., $x$ is considered a typical situation of $p$ even though $p$ is not true at $x$ in $M_\alpha^R$.

## 5.3 Unified Formulation of Reasoning

In this section, we characterize the reasoning processes of deduction, induction, and abduction in $\alpha$-level fuzzy measure models on the basis of background knowledge as reasoning processes based on typical situations.

Now, we treat deduction based on typical situations. Deduction is a reasoning process with the following form:

| | |
|---|---|
| $p \to q$ | If $P$, then $Q$. |
| $p$ | $P$. |
| $q$ | Therefore, $Q$. |

where the left side illustrates the formulation of deduction, and the right side illustrates the meaning of the sentence appearing in each deductive step. It is well known that deduction is identical to the inference rule *modus ponens* used in almost all two-valued logic.

Note also that deduction is a logically valid inference, where "logically valid" means that if both the antecedent $p$ and the rule $p \to q$ are true, the consequent $q$ is guaranteed to be true. Hereafter, we assume that all sentences $p, q$, etc. that represent facts and rules such as $p \to q$ are non-modal sentences.

Let $M = (U, R, v)$ be a Kripke model that consists of an approximation space $(U, R)$ and a valuation function $v$ that is given as background knowledge. In the framework of possible world semantics, we can illustrate deduction as follows:

| | |
|---|---|
| $M \models p \to q$ | (In any situation) If $P$, then $Q$. |
| $M, x \models p$ | (In a situation) $P$. |
| $M, x \models q$ | (In the situation) $Q$. |

Here, we consider deduction based on typical situations. Let $M_\alpha^R$ be an $\alpha$-level fuzzy measure model based on background knowledge with a fixed degree $\alpha \in (0.5, 1]$.

True rules are represented by inclusion relationships between truth sets as follows:

$$M_\alpha^R \models p \to q \Leftrightarrow \|p\|^{M_\alpha^R} \subseteq \|q\|^{M_\alpha^R}. \tag{5.1}$$

As the monotonicity of $\beta$-lower approximation is satisfied for all $\beta \in [0, 0.5)$; thus, we have the relationship,

$$M_\alpha^R \models \Box p \to \Box q \Leftrightarrow \|\Box p\|^{M_\alpha^R} \subseteq \|\Box q\|^{M_\alpha^R}. \tag{5.2}$$

If we consider the truth set of $\Box p$ as the set of typical situations of $p$, then from (5.2), every element $x \in \|\Box p\|^{M_\alpha^R}$ is also an element in the truth set of $\Box q$, and therefore, we can conclude that all situations typical of $p$ are also typical of $q$.

Consequently, using the $\alpha$-level fuzzy measure model $M_\alpha^R$, we can characterize deduction based on typical situations by the following valid reasoning:

| | |
|---|---|
| $M_\alpha^R \models \Box p \to \Box q$ | If (typically) $P$, then (typically) $Q$. |
| $M_\alpha^R, x \models \Box p$ | (Typically) $P$. |
| $M_\alpha^R, x \models \Box q$ | (Typically) $Q$. |

Note that the reasoning process of deduction based on typical situations is not affected by a difference in the degree $\alpha$. This is because property (5.2) is true for any fixed degree $\alpha \in (0.5, 1]$, and therefore, if a possible world $x$ is a typical situation of a fact $p$ and a modal sentence $\Box p \to \Box q$ is valid in the $\alpha$-level fuzzy measure model $M_\alpha^R$, then $x$ is also a typical situation of the fact $q$.

As an example of deduction, suppose sentences $p$ and $q$ have the following meanings:

- $p$: The sun rises in the east.
- $q$: The sun sets in the west.

Thus, deduction is illustrated as follows:

| | |
|---|---|
| $M_\alpha^R \models \Box p \to \Box q$ | If the sun rises in the east, then the sun sets in the west. |
| $M_\alpha^R, x \models \Box p$ | Today, the sun rises in the east. |
| $M_\alpha^R, x \models \Box q$ | The sun will set in the west today. |

We now turn to induction based on typical situations. Induction is a reasoning process with the following form:

| | |
|---|---|
| $p$ | $P$. |
| $q$ | $Q$. |
| $p \to q$. | Therefore, if $P$, then $Q$. |

It is well known that induction is not logically valid. However, we often use induction to provide general rules from specific facts.

Induction has the following characteristic: From the fact that all observed objects satisfying a property $p$ also satisfy a property $q$, we conclude that if objects satisfy $p$, they also satisfy $q$. Suppose that the $(1-\alpha)$-lower approximation of the truth set $\|p\|^{M_\alpha^R}$ of the sentence $p$ illustrates the set of observed objects satisfying $p$. From the characteristics of induction, we consider that induction based on typical situations needs to have the form:

| | |
|---|---|
| $M_\alpha^R \models \Box p \to q$ | If observed objects satisfy $P$, then the objects also satisfy $Q$. |
| $M_\alpha^R \models p \to q$ | If $P$, then $Q$. |

This form of reasoning is not valid; however, we can consider this reasoning as valid by assuming the property:

$$M_\alpha^R \models \Box p \leftrightarrow p. \tag{5.3}$$

This assumption means that we consider the set $\|\Box p\|^{M_\alpha^R}$ of observed objects satisfying $p$ is identical to the set $\|p\|^{M_\alpha^R}$ of all objects satisfying $p$; i.e., we generalize from the typical situations of $p$ to all situations of $p$. This assumption is essential in formulating induction based on typical situations. Combining these processes of reasoning, we characterize induction based on typical situations as follows:

| | |
|---|---|
| $M_\alpha^R \models \Box p \to q$ | If observed objects satisfy $P$, then the objects also satisfy $Q$. |
| $M_\alpha^R \models \Box \leftrightarrow p$ | Generalization of observation. |
| $M_\alpha^R \models p \to q$ | If $P$, then $Q$. |

By repeating observations, we obtain more detailed background knowledge, and assumption (5.3) may become more probable. As shown in Table 5.1, in VPRS models, even though the partition becomes finer (that is, the current equivalence relation $R$ changes to another equivalence relation $R'$ such that $R' \subseteq R$, the $\beta$-lower approximation may not become large.

However, the following situation may result from the more detailed equivalence relation $R'$:

For any $q$, $M_\alpha^R \models \Box p \to q$ but $M_\alpha^{R'} \not\models \Box p \to q$

This situation illustrates that by obtaining more detailed background knowledge, we find exceptions in the observed objects such that they do not satisfy q even while satisfying $p$. Therefore, in the framework of the $\alpha$-level fuzzy measure model based on background knowledge, induction has *non-monotonicity*.

This consideration indicates that, unlike deduction based on typical situations, the degree of $\alpha \in (0.5, 1]$ may affect the result of induction based on typical situations; assumption (5.3) with $\alpha = 1$ may be more reliable than the assumption with $\alpha \in (0.5, 1)$. This is because if $\alpha = 1$ then the modal sentence $\Box p \to p$ is valid in any $\alpha$-level fuzzy measure model $M_\alpha^R$ based on background knowledge.

On the other hand, if $\alpha \in (0.5, 1)$, this modal sentence may be not true in some observed object $x \in \|\Box p\|^{M_\alpha^R}$, and such an object $x$ becomes a counterexample of the assumption.

As an example of induction and nonmonotonic reasoning, suppose sentences $p$ and $q$ have the following meanings:

- $p$: It is a bird.
- $q$: It can fly.

Thus, induction and non-monotonic reasoning are illustrated as follows:

| | |
|---|---|
| $M_\alpha^R \models \Box p \to q$ | All observed birds can fly. |
| $M_\alpha^R \models \Box p \leftrightarrow p$ | Generalization of observation. |
| $M_\alpha^R \models p \to q$ | Therefore, all birds can fly. |

The equivalence relation $R$ changes to a more detailed equivalence relation $R'$ by repeating observations.

| | |
|---|---|
| $M_\alpha^R \not\models \Box p \to q$ | Not all birds can fly. |

Next, we discuss abduction which is a reasoning process with the following form:

| | |
|---|---|
| $q$ | $Q$, |
| $p \to q$ | If $P$, then $Q$, |
| $p$ | Therefore, $P$. |

From a fact $q$ and a rule $p \to q$, abduction infers a hypothesis $p$ that produces the fact $q$. Therefore, abduction is also called *hypothesis reasoning*.

Note that the form of abduction corresponds to affirming the consequent; thus, abduction is not logically valid if the hypothesis $p$ is false and the fact $q$ is true. However, we often use this form of reasoning to generate new ideas.

In general, many rules may exist that produce the fact $q$, and in such cases, we need to select one rule from many $p_i \to q$ $(p_i \in \{p_1, ..., p_n\})$ that imply $q$. Thus, using fuzzy measures assigned to typical situations of the fact $q$, we introduce a selection mechanism to decide which rule to use in abduction.

Similar to the case of deduction, we consider the truth set $\|\Box q\|^{M_\alpha^R}$ of $\Box q$ as the set of typical situations about $q$. For each rule $p_i \to q$ that implies the fact $q$, we consider the following minimal degree of the antecedent $p_i$ in typical situations about $q$.

**Definition 5.2** Let $p \to q, q \in L_{ML}(\mathscr{P})$ be non-modal sentences. Then, degree $\alpha(p \mid q)$ of $p$ in typical situations about $q$ is defined as follows:

$$\alpha(p \mid q) = \begin{cases} \min\{\mu_x^R(\|p\|^{M_\alpha^R}) \mid x \in \|\Box q\|^{M_\alpha^R}\}, & \text{if } \|\Box q\|^{M_\alpha^R} \neq \emptyset, \\ 0, & \text{otherwise.} \end{cases}$$

To demonstrate the calculation of the degree $\alpha(p \mid q)$, we present an example. Let $M = (U, R, v)$ be a Kripke model that consists of the set of possible worlds

$U = \{w_1, ..., w_{10}\}$, an equivalence relation $R$, and a valuation function $v$. The equivalence relation $R$ provides the following three equivalence classes:
$[w_1]_R = \{w_1, w_2, w_3\}$,
$[w_4]_R = \{w_4, w_5, w_6, w_7\}$,
$[w_8]_R = \{w_8, w_9, w_{10}\}$.

Moreover, the truth sets of three non-modal sentences $p_1$, $p_2$, and $q$ in $M$ are:
$\|p_1\|^M = \{w_1, w_2, w_3, w_4, w_5, w_6\}$,
$\|p_2\|^M = \{w_2, w_3, w_4, w_5, w_6, w_7\}$,
$\|q\|^M = \{w_1, w_2, w_3, w_4, w_5, w_6, w_7, w_8\}$.
Note that both $p_1$ and $p_2$ conclude $q$.

Suppose we fix $\alpha = 0.7$, and consider the $\alpha$-level fuzzy measure mode $M_\alpha^R$ a based on background knowledge $M$. Here, for the two rules $p_1 \rightarrow q$ and $p_2 \rightarrow q$, we calculate the degrees $\alpha(p_1 \mid q)$ and $\alpha(p_2 \mid q)$, respectively. The set of typical situations of $q$ in $M_\alpha^R$ is the set

$$\|\Box q\| = [w_1]_R \cup [w_4]_R = \{w_1, w_2, w_3, w_4, w_5, w_6, w_7\}$$

For $\alpha(p_1 \mid q)$, we need to calculate the degrees of the truth set $\|p\|^M$ by the fuzzy measures $\mu_x^R$ as follows:

$$\mu_{w_i}^R(\|p_1\|^M) = \frac{|[w_1]_R \cap \|p_1\|^M|}{|[w_1]_R|} = \frac{|\{w_1, w_2, w_3\}|}{|\{w_1, w_2, w_3\}|} = 1, \quad w_i \in [w_1]_R,$$

$$\mu_{w_j}^R(\|p_1\|^M) = \frac{|[w_4]_R \cap \|p_1\|^M|}{|[w_1]_R|} = \frac{|\{w_4, w_5, w_6\}|}{|\{w_4, w_5, w_6, w_7\}|} = \frac{3}{4}, \quad w_j \in [w_4]_R$$

Thus, we have the degree $\alpha(p_1 \mid q)$ as follows:

$$\alpha(p_1 \mid q) = \min\left\{1, \frac{3}{4}\right\} = \frac{3}{4}.$$

Similarly, we also calculate the degree $\alpha(p_2 \mid q) = \frac{2}{3}$.

For any non-modal sentence $p \rightarrow q$, the degree $\alpha(p \mid q)$ satisfies the following good property.

**Proposition 5.1** *Let $p, q \in L_{ML}(\mathscr{P})$ be a non-modal sentences. For any $\alpha$-level fuzzy measure model based on background knowledge $M_\alpha^R$ with the fixed degree $\alpha \in (0.5, 1]$, if the condition $\|\Box q\|^{M_\alpha^R} \neq \emptyset$ holds, the following property is satisfied:*

$$M_\alpha^R \models \Box q \rightarrow \Box p \Leftrightarrow \alpha(p \mid q) \geq \alpha.$$

*Proof* ($\Leftarrow$): Suppose that $\alpha(p \mid q) \geq \alpha$ holds. Because we have $|\Box q\|^{M_\alpha^R} \neq \emptyset$ by the assumption of the proposition, there is a possible world $y \in \|\Box q\|^{M_\alpha^R}$ such that $\alpha(p \mid q) = \mu_y^R(\|p\|^{M_\alpha^R})$ and $\mu_y^R(\|p\|^{M_\alpha^R}) \leq \mu_x^R(\|p\|^{M_\alpha^R})$ for all typical situations $x \in \|\Box q\|^{M_\alpha^R}$. Because $\alpha(p \mid q) \geq \alpha$ holds, $\mu_x^R(\|p\| \,|^{M_\alpha^R}) \geq \alpha$ for all $x \in \|\Box q\|^{M_\alpha^R}$. Therefore, we have $\|\Box q\|^{M_\alpha^R} \subseteq \|\Box p\|^{M_\alpha^R}$, which leads to $M_\alpha^R \models \Box q \rightarrow \Box p$.

($\Rightarrow$): Suppose that $M_\alpha^R \models \Box q \rightarrow \Box p$ holds. This property implies that $\|\Box q\|^{M_\alpha^R} \subseteq \|\Box p\|^{M_\alpha^R}$. Moreover, because we have $\|\Box q\|^{M_\alpha^R} \neq \emptyset$ by assumption, at least one typical situation of $q$ exists. Thus, for all typical situations $x \in \|\Box q\|^{M_\alpha^R}$ of $q$,
$\mu_x^R(\|p\|^{M_\alpha^R}) \geq \alpha$ holds. Therefore, by the definition of the degree $\alpha(p \mid q)$, we conclude that $\alpha(p \mid q) \geq \alpha$ holds.

Proposition 5.1 indicates that we can use the degree $\alpha(p \mid q)$ as a criterion to select a rule $p \to q$ that implies the fact $q$. For example, from many rules $p_i \to q$ ($p_i \in \{p_1, ..., p_n\}$) that imply $q$, we can select a rule $p_j \to q$ with the highest degree that $\alpha(p_j \mid q)$ such that $\alpha(p_j \mid q) \geq \alpha$.

In this case, we consider the selected rule $p_j \to q$ as the most universal rule to explain the fact $q$ in the sense that all typical situations of $q$ fit the typical situations of $p_j$. Thus, in the above example, we select the rule $p_1 \to q$ because we have $\alpha(p_1 \mid q) \geq \alpha = 0.7$ but $\alpha(p_2 \mid q) < \alpha$.

On the other hand, we can consider the case that no rule satisfies Definition 5.2 as a situation in which we cannot explain the fact $q$ by the current background knowledge.

Therefore, by selecting the rule $p \to q$ with the highest degree $\alpha(p \mid q)$ such that $\alpha(p \mid q) \geq \alpha$, we can characterize abduction that infers $p$ from the fact $q$ based on typical situations by the following form of valid reasoning:

| | |
|---|---|
| $M_\alpha^R, x \models \Box q$ | (Actually) $Q$, |
| $M_\alpha^R \models \Box q \to \Box p$ | Selection of a rule "if $P$, then $Q$, |
| $M_\alpha^R, x \models \Box p$ | (Perhaps) $P$. |

By this formulation of abduction based on typical situations, it is clear that the difference of the degree $\alpha \in (0.5, 1]$ affects the result of abduction.

As an example of abduction (or hypothesis reasoning), we consider reasoning based on fortune-telling. Suppose sentences $p_1$, $p_2$, and $q$ used in the above example have the following meanings:

- $p_1$: I wear some red items.
- $p_2$: My blood type is AB.
- $q$: I am lucky.

Then, using the $\alpha$-level fuzzy measure model $M_\alpha^R$ based on background knowledge $M$ in the above example, reasoning based on fortune-telling is characterized by abduction as follows:

| | |
|---|---|
| $M_\alpha^R, x \models \Box q$ | I am very lucky today! |
| $M_\alpha^R \models \Box q \to \Box p$ | In a magazine, I saw that "wearing red items makes you lucky" |
| $M_\alpha^R, x \models \Box p$ | Actually I wear red socks! |

In this section, we have introduced an $\alpha$-level fuzzy measure model based on background knowledge and proposed a unified formulation of deduction, induction, and abduction based on this model.

Using the proposed model, we have characterized typical situations of the given facts and rules by $(1-\alpha)$-lower approximation of truth sets of nonmodal sentences that represent the given facts and rules.

We have also proven that the system EMND45 is sound with respect to the class of all $\alpha$-level fuzzy measure models based on background knowledge. Moreover, we have characterized deduction, induction, and abduction as reasoning processes based on typical situations.

In the proposed framework, deduction and abduction are illustrated as valid reasoning processes based on typical situations of facts. On the other hand, induction is illustrated as a reasoning process of generalization based on observations.

Furthermore, in the $\alpha$-level fuzzy measure model based on background knowledge, we have pointed out that induction has non-monotonicity based on revision of the indiscernibility relation in the given Kripke model as background knowledge and gave an example in which a rule inferred by induction based on typical situations is rejected by refinement of the indiscernibility relation.

## 5.4 Non-monotonic Reasoning

Common-sense reasoning is not generally monotonic. This means that new information cannot invalidate old conclusions and that classical logic is not adequate to formalize common-sense reasoning. Minsky [3] criticized classical logic as a knowledge representation language. Here, some words may be in order.

Let $\Gamma$, $\Gamma'$ be a set of formulas and $A$, $B$ be formulas. In classical logic **CL**, if we can prove $A$ from $\Gamma$, then $A$ is seen as a theorem. **CL** is monotonic, i.e., if $\Gamma \subseteq \Gamma'$ then $Th(\Gamma) \subseteq Th(\Gamma')$, where $Th(\Gamma) = \{B \mid \Gamma \vdash_{\mathbf{CL}} B\}$.

Common-sense reasoning is, however, not monotonic, as Minsky observed. We may revise old conclusions when new information is available. But, inconsistency arises in such a case since classical logic is monotonic.

Now, consider the famous Penguin example.

(1) All birds fly.
(2) Tweety is a bird.

(1) and (2) are represented as formulas in **CL** as follows:

(3) $\forall x(bird(x) \rightarrow fly(x))$
(4) $bird(tweety)$

From (3) and (4), we have:

(5) $fly(tweety)$

Namely, we can conclude that Tweety flies. But, assume that new information "Tweety is a penguin". Consequently, the following two formulas are added to the database:

(6) $penguin(tweety)$
(7) $\forall x(penguin(x) \rightarrow \neg fly(x))$

From (6) and (7), (8) can be derived:

(8) $\neg fly(tweety)$

This is a new conclusion. If our underlying logic is classical logic, then (5) and (8) are contradictory. In classical logic, everything can be derived, and we cannot obtain correct information about a penguin. On the other hand, in non-monotonic logics, we think that non-monotonic reasoning, i.e. from (5) to (8), happens.

Since the 1980s, AI researchers have proposed the so-called *non-monotonic logic* which is a logical system capable of formalizing non-monotonic reasoning. There are several non-monotonic logics, which include the non-monotonic logic of McDermott and Doyle, *autoepistemic logic* of Moore, *default logic* of Reiter, and *circumscription* of McCarthy. Below, we concisely survey major non-monotonic logics.

First, we review McDermott and Doyle's non-monotonic logic $NML1$; see McDermott and Doyle [4]. The language of $NML1$ extends the language of first-order classical logic with the *consistency operator*, denoted M. The formula M$A$ reads "$A$ is consistent". The basic idea of $NML1$ is to formalize non-monotonic logic as a modal logic. In this sense, non-monotonicity can be expressed at the object-level.

In $NML1$, (1) is represented as follows:

(9) $\forall x(bird(x) \& \mathrm{M}(fly(x)) \rightarrow fly(x))$

The interpretation of (9) is that $x$ is a bird and we have no counter-examples to flying, $x$ flies. Here, the consistency of $fly(x)$ means the underivability of $\neg fly(x)$.

Thus, if we have the information "Tweety is a penguin", then we have $\neg fly(tweety)$. Thus, (1) cannot be applied. And the conclusion that Tweety flies can be withdrawn due to new informarion.

In $NML1$, non-monotonic inference relation $\mid\sim$ is defined. $T \mid\sim A$ means that a formula $A$ is non-monotonically proved from the theory $T$. However, the definition is circular, and McDermott and Doyle defined $\mid\sim$ by means of fixpoint as follows:

$$T \mid\sim A \;\Leftrightarrow\; A \in Th(T) = \cap\{S \mid S = NM_T(S)\}$$

Here, the existence of the fixpoint of $NM_T$ is assumed. When there are no fixpoints of $NM_T$, $Th(T) = T$. We define $NM_T$ as follows:

$$NM_T(S) = Th(T \cup AS_T(S))$$
$$AS_T(S) = \{MA \mid A \in L \text{ and } \neg A \notin S\} - Th(T)$$

Although McDermott and Doyle defined fixpoint in this way, there are several difficulties. Indeed, the definition is mathematically strict, but it is not compatible to our intuition. In addition, there are cases that several fixpoints exist. On these grounds, the formalization of non-monotonic reasoning in $NML1$ is not appropriate.

To overcome such defects in $NML1$, McDermott proposed $NML2$ in [5]. The essential difference of $NML1$ and $NML2$ is that the latter is based on modal logic and the former is based on classical logic. Thus, $NML2$ describes non-monotonic reasoning in terms of Kripke semantics for modal logic. Therefore, $NML2$ uses $Th_X(T)$ instead of $Th(T)$:

$$Th_X(T) = \cap\{S \mid S = X - NM_T(S)\}.$$

McDermott introduced a noncommittal model and proved that $T \mid\!\sim_X A$ iff $A$ is true for every $X$-noncommittal model in which $T$ is true, where $X = \{\mathbf{T}, \mathbf{S4}, \mathbf{S5}\}$. Unfortunately, the relation cannot capture the intuitive meaning of non-monotonic reasoning. Further, as McDermott proved, monotonic and non-monotonic S5 agree.

Thus, one considered that non-monotonic modal logics have some limitations for modeling non-monotonic reasoning. Later, the defects have been overcome by Moore's autoepistemic logic, and we saw some theoretical developments for non-monotonic modal logics themselves; see Marek, Shvarts and Truszczynski [6].

Reiter proposed *default logic* in Reiter [7]. Its aim is to describe default reasoning which is a reasoning by default in the lack of counter-examples. Default reasoning is expressed by means of the inference rule called *default.*

Default logic is an extension of classical first-order logic and default is a meta-level concept. This means that non-monotonic reasoning cannot be formalized in the object-level unlike in $NML1$ and $NML2$.

A *default theory* is expressed as a pair $\langle D, W \rangle$, where $W$ is a set of formulas of first-order logic (i.e., theory) and $D$ is a default of the form:

$$(1) \quad \frac{A : B_1, ..., B_n}{C}$$

where $A$ is a *prerequisite* (prerequisite), $B_1, ..., B_n$ are a *justification*, and $C$ is a *consequent*, respectively. We call a default theory containing free variables a *open default* and a default not containing free variables *closed default*, respectively.

The default of the form (2) is called *normal default* and (3) *semi-normal default*, respectively.

$$(2) \quad \frac{A : B}{B}$$

$$(3) \quad \frac{A : B \& C}{C}$$

Default (1) reads "If $A$ is provable and $B_1, ..., B_n$ are all consistent, then we may conclude $C$". Thus, the above penguin example is described as follows:

$$\frac{bird(x) : fly(x)}{fly(x)}$$

Reiter introduced the concept of *extension* to describe default reasoning. An extension of the default theory $\delta = \langle D, W \rangle$ with a set $E$ of formulas of first-order logic is defined as a least set satisfying the following conditions:

(1) $W \subseteq \Gamma(E)$
(2) $Th(\Gamma(E)) = \Gamma(E)$
(3) $(A : B_1, ..., B_n/C) \in D$ and $A \in \Gamma(E), \neg B_1, ..., \neg B_n \notin E \Rightarrow C \in \Gamma(E)$.

The extension of $\delta$ is defined as a fixpoint $E$ of $\Gamma$, i.e., $\Gamma(E) = E$. Therefore, default reasoning is expressed by means of the notion of extension. The extension for a default theory does not always exist, but Reiter proved that for normal default theory at least one extension exists; see [7].

A proof theory for default logic is given by a *default proof*, which is different from the one in standard logical systems. In other words, it provides a procedure for proving whether certain formula is included in an extension.

Let $\delta = \langle D, W \rangle$ be a normal default and $A$ be a formula. A default proof for $\delta$ is a finite sequence $D_0, D_1, ..., D_n$ each of which is a finite subset of $D$, satisfying the following properties:

(1) $W \cup CONS(D_0) \vdash A$
(2) for every $F \in PRE(D_{i-1})$, $W \cup CONS(D_i) \vdash F$ $(1 \leq i \leq n)$
(3) $D_n = \emptyset$
(4) $\bigcup_{i=0}^{n} CONS(D_i)$ is consistent with $W$.

Here, $PRE(D_i)$ is a prerequisite, $CONS(D_i)$ is a consequent, respectively. Reiter showed that linear resolution can serve to give a default proof.

Etherington proposed a model theory for default logic in [8]. It is, however, pointed out that it is too complicate to capture the intuitive meaning of default reasoning. Lukasiewicz developed a version of default logic to guarantee the existence of extension as an improvement of Reiter's default logic in Lukasiewicz [9, 10]. For details on default logic, see Etherington [8] and Besnar [11].

*Autoepistemic logic* was proposed by Moore [12, 13] as an alternative to McDermott and Doyle's non-monotonic logic $NML1$. It is a non-monotonic logic to model beliefs of a *rational agent* who can reflect by himself. An agent's beliefs always change as the information he can obtain increases, and consequently autoepistemic logic is a non-monotonic logic. However, *autoepistemic reasoning* should be distinguished by default reasoning performed by the lack of information.

Autoepistemic logic is an extension of classical propositional logic with the modal operator L. A formula L$A$ reads "$A$ is believed by the agent". Thus, L corresponds to the belief operator Bel (believe) in belief logic (cf. Hintikka [14]). Now, consider how the beliefs of a rational agent are formalized. The point is that he should infer about what he believes and what he does not believe.

In 1980, Stalnaker proposed the conditions the rational agent's beliefs satisfy in Stalnaker [15].

(1) $P_1, ..., P_n \in T$ and $P_1, ..., P_n \vdash Q \Rightarrow Q \in T$
(2) $P \in T \Rightarrow \mathrm{L}P \in T$
(3) $P \notin T \Rightarrow \neg\mathrm{L}P \in T$

An autoepistemic theory which is a set of formulas is called *stable* if it satisfies the above three conditions.

To give a semantics for autoepistemic logic, Moore later introduced *autoepistemic interpretation* and *autoepistemic model* in Morre [12]. An autoepistemic interpretation $I$ for autoepistemic theory $T$ is a truth assignment to a formula satisfying the following conditions:

(1) $I$ satisfies the truth conditions of propositional logic $PL$
(2) L$A$ is true in $I$ $\Leftrightarrow A \in T$.

An autoepistemic model for autoepistemic theory $T$ is the autoepistemic interpretation satisfying that all the formulas in $T$ are true. $T$ is *semantically complete* iff $T$ contains every formulas that is true in every model.

Moore proved that the concept stability characterizes semantically complete autoepistemic theory. He also introduced the notion of *groundedness*, which corresponds to soundness of autoepistemic logic.

An autoepistemic theory is grounded for an agent's assumption $S$, iff all the formulas of $S$ are contained in the logical consequence of (4):

(4) $S \cup \{\mathrm{L}A \mid A \in T\} \cup \{\neg \mathrm{L}A \mid A \notin T\}$

An autoepistemic theory $T$ is sound if it is grounded in $S$. The beliefs of a rational agent is both semantically complete and grounded with respect to his beliefs. An autoepistemic theory $T$ is a *stable expansion* of the assumption $P$, iff $T$ is a logical consequence of (5):

(5) $P \cup \{\mathrm{L}A \mid A \in T\} \cup \{\neg \mathrm{L}A \mid A \notin T\}$

Here, $P$ is a set of formulas not containing L and $A$ is a formulas not containing L.

Although autoepistemic and default logics are essentially different, there are connections by translations as Konolige [16] showed.

Autoepistemic model is considered to be a self-referential model, but Moore [12] proposed a Kripke type possible world model for it. It is not surprising that there is a possible world semantics for autoepistemic logic. This is because a possible world in a Kripke model is a world an agent believes like the actual world. Namely, an agent believes $A$ when it is true in a world believed to be the actual world by him.

It is known that the accessibility relation of Kripke model for modal logic S5 is an equivalence relation. We call S5 model in which all worlds are accessible from all worlds *complete S5 model*. Moore proved that a set of formulas which are true in every complete S5 model coincides with stable autoepistemic theory.

Let $K$ be complete S5 model and $V$ be truth assignment. Then, $(K, V)$ is a possible world interpretation of $T$ iff $T$ is a set of formulas which are true in every world in $K$. We define that $(K, V)$ is a possible world model for $T$ if every formula in $T$ is true in $(K, V)$. Moore proved that there is a possible world model for autoepistemic model of stable theory (and its converse).

Moore's possible world model is an alternative formulation of autoepistemic model, but in Kripke model for belief logic the belief operator is interpreted as the necessity operator. In this sense, Moore's possible world model is different from standard Kripke model.

It is thus interesting to consider whether the operator L can be directly in a Kripke model. This is to identify a modal system underlying autoepistemic logic. As Stalnaker pointed out, autoepsitemic logic corresponds to *weak S5*. It is exactly the modal system *KD45*. If we allow that an agent's beliefs are inconsistent, then the corresponding modal system is $K45$.

It is difficult to formalize L in autoepistemic since it is non-monotonic. If we interpret autoepistemic reasoning in monotonic modal logic, difficulties with Moore's model can be avoided. Such an attempt was done by Levesque [17] who introduced a logic of All I know, denoted $OL$. $OL$ extends first-order logic with the three modal operators L, N and O. O$A$ reads "All I know is $A$", N$A$ reads "At least I know is $A$", and L is the one in Moore's autoepistemic logic, respectively. O$A$ is defined as L$A$&N$\neg A$.

Levesque axiomatized O instead of L to formalize autoepistemic reasoning. The reason is that O can be formalized in monotonic modal logic. The Hilbert system for $OL$ extends first-order logic with the following axioms:

(OL1) $*A$ ($A$ is a formula not containing modal operators)
(OL2) $*(A \rightarrow B) \rightarrow (*A \rightarrow *B)$
(OL3) $\forall x * A(x) \rightarrow *\forall x A(x)$
(OL4) $A \rightarrow *A$ (All predicates in $A$ are in the scope of modal operators)
(OL5) $\mathrm{N}A \rightarrow \neg \mathrm{L}A$ ($\neg A$ is a satisfiable formula not containing modal operators)

Here, inference rules are the same as in first-order logic. The symbol $*$ denotes a modal operator L or N.

The semantics for $OL$ is given by a Kripke model $\langle W, R, V \rangle$ for $KD45$. The truth relation $W, w \models A$ is defined as $w(A) = 1$ for all atomic formulas $A$. A set of possible worlds is assumed to be a set of truth assignments $w$. Thus, formulas L$A$ and N$A$ are interpreted as follows:

(1) $W, w \models \mathrm{L}A \Leftrightarrow \forall w' \in W, W, w' \models A$
(2) $W, w \models \mathrm{N}A \Leftrightarrow \forall w' \notin W, W, w' \models A$

From (1) and (2), we have the interpretation of O$A$:

(3) $W, w \models \mathrm{O}A \Leftrightarrow W, w \models \mathrm{L}A$ and $\forall w' \in W(W, w' \models A \Rightarrow w' \in W)$

Levesque proved that the propositional fragment of $OL$ is complete for the Kripke semantics.

In $OL$, non-monotonic inference relation $\mid\!\sim$ is defined at the object-level as follows:

(4) $P \mid\!\sim A \Leftrightarrow \mathrm{O}P \rightarrow \mathrm{L}A$

The major advantage of $OL$ is that non-monotonic reasoning can be described in the object-language. For example, the lack of extension is written as $\neg\mathrm{O}(\neg \mathrm{L}A \rightarrow \neg A)$, which is a theorem of $OL$.

The above non-monotonic logics extend classical logic to formalizing non-monotonic reasoning. There are, however, non-monotonic theories within the framework of classical logic. *Circumscrption* is one of such non-monotonic theories.

Circumscription was introduced by McCarthy. There are two sorts of circumscription. One is *predicate circumscription* (cf. McCarthy [18]) and *formula circumscription* (cf. McCarthy [19]).

In 1980, McCarthy proposed predicate circumscription which can minimize certain predicate. Let $A$ be a first-order formula containing a predicate $P(x)$ and $A(\Phi)$ be the formula obtainable from $A$ by substituting predicate expression $\Phi$ for all $P$ in $A$. Here, a predicate $P$ is $n$-place predicate.

Predicate circumscription of $A(P)$ is defined as the following second-order formula:

$$A(\Phi)\&\forall x(\Phi(x) \rightarrow P(x)) \rightarrow \forall x(P(x) \rightarrow \Phi(x))$$

Here, we describe some inferences in the block world by using circumscription. Assume that a block world is given as follows:

(1) $is\_block(A)\&is\_block(B)\&is\_block(C)$

(1) states that $A$, $B$ and $C$ are a block. If we minimiz the predicate $is\_block$ in (1), then the following holds.

(2) $\Phi(A)\&\Phi(B)\&\Phi(C)\&\forall x(\Phi(x) \rightarrow is\_block(x)) \rightarrow \forall x(is\_block(x) \rightarrow \Phi(x))$

Here, if we substitute (3) for (2) and use (1), then we have (4):

(3) $\Phi(x) \leftrightarrow (x = A \vee x = B \vee x = C)$
(4) $\forall x(is\_block(x) \rightarrow (x = A \vee x = B \vee x = C))$

Semantics for predicate circumscription is based on *minimal model*. And *minimal entailment* is defined by means of minimal model. A formula $q$ is minimally entailed by a formula $A$ iff $q$ is truc in all minimal models for $A$. Therefore, minimal entailment is seen as a model-theoretic interpretation of predicate circumscription.

Here, we precisely define the concept of minimal model. Let $M$, $N$ be models of a formula $A$. If domains of $M$ and $N$ are equal, and the extensions (interpretations) of predicates except $P$ are equal, and the extension of $P$ in $M$ is included in the extension of $P$ in $N$, then $M$ is called a *submodel* of $N$, written $M \leq_P N$. We say that model $M$ is $P$-minimal model of $A$ iff every model $M'$ of $A$ which is $P$-minimal model of $M$ is equal to $M$. Namely,

$$M' \leq_P M \Rightarrow M' = M.$$

$A$ minimally entails $q$ with respect to $P$ iff $q$ is true in every $P$-minimal model of $A$, written $A \models_P q$.

McCarthy proved soundness of predicate circumscription for the above semantics. Namely,

$$A \vdash_{circ} q \Rightarrow A \models_P q$$

where $\vdash_{circ}$ denotes the inference relation of circumscription. But, it is known that its converse, i.e., completeness does not in general hold.

Predicate circumscription minimizes certain predicate by fixing other predicates. Thus, it cannot generally deal with non-monotonic reasoning, since the extensions of various predicates vary according to the increase of information.

For instance, even if we minimize exceptional birds (e.g. penguin), a set of flying objects may increase. Consequently, the whole formulas in a database should be minimized for non-monotonic reasoning.

In 1986, McCarthy proposed *formula circumscription* to overcome the defect of predicate circumscription in [19]. It can minimize the whole formula and uses a special predicate *abnormal* to handle exceptions. Formula circumscription is defined as the following second-order formula:

$$Circum(A; P; Z) = A(P, Z) \ \& \ \neg\exists pz(A(p, z) \ \& \ p < P).$$

where $P$ denotes a tuple of predicate constants, $Z$ a tuple of predicate constants not occurring in $P$, and $A(P, Z)$ a formula in which the elements of $P, Z$ occur.

There are two differences of predicate and formula circumscriptions. First, the former speaks of a specific predicate, whereas the latter speaks of specific formula. However, formula circumscription is reducible to predicate circumscription, provided that we allow as variables predicates besides the one being minimized.

Second, in the above definition of formula circumscription, we use an explicit quantifier for the predicate variable, whereas in predicate circumscription the formula is a scheme.

By this definition, formula circumscription yields minimal interpretation of certain predicate and the predicates which is false in the interpretation means negative information. In formula circumscription, the penguin example is described as follows:

(1) $\forall x((bird(x) \ \& \ \neg ab_1(x)) \rightarrow fly(x))$
(2) $bird(tweety)$

Here, (1) reads "Birds fly unless it is not an exception ($ab_1$)". Now, we write $A_1$ for conjunction of (1) and (2). Then, formula circumscription can give an interpretation of minimizing the exception ($ab_1$) about flying.

Applying formula circumscription to (1), we have (3):

(3) $A_1(ab_1, fly) \ \& \ \neg\exists p_1 z(A_1(p_1, z) \ \& \ p_1 < ab_1)$

In (3), we substitute $ab_1$ and $fly$ for $p_1$ and $z$, respectively, and (4) is obtained:

(4) $A_1(ab1, fly) \ \& \ \neg\exists p_1 z(\forall x(bird(x) \ \& \ \neg p_1(x) \rightarrow z(x)) \ \& \ bird(tweety)$
$\& \ \forall x(p_1(x) \rightarrow ab_1(x)) \ \& \ \neg\forall x(ab_1(x) \leftrightarrow p_1(x)).$

Next, we substitute $false$ for $p_1$ and $true$ for $z$, respectively. Then, we have (5):

(5) $A_1(ab_1, fly) \ \& \ \neg p_1 z(\forall x(bird(x) \ \& \ \neg false \rightarrow true) \ \& \ bird(tweety)$
$\& \ \forall x(false \rightarrow ab_1(x)) \ \& \ \neg\forall x(ab_1(x) \leftrightarrow false))$

If $\neg\forall x(ab_1(x) \leftrightarrow false)$ is true, there there exist predicates $p_1$ and $z$, making the second conjunct of (3) true. However, (3) intends that there are no such $p_1, z$. Thus, (6) holds:

(6) $\forall x(ab_1(x) \leftrightarrow false)$

which is equivalent to (7).

(7) $\forall x \neg ab_1(x)$

(7) is a conclusion which is obtainable from $Circum(A_1; ab_1; fly)$. Namely, from (1) and (2), it follows that there are no exceptions. This is obvious since we do not have the description of exceptions

Next, we explain the representation of non-monotonic reasoning in formula circumscription. Assume that the new information (8) is added to the database $A_1$.

(8) $\neg fly(tweety)$

Let the new database $A_1$ & (8) be $A_2$. Then, $Circum(A_2; ab_1; fly)$ becomes the following:

$$(9)\ A_2(ab_1, fly) \,\&\, \neg\exists p_1 z(\forall x(bird(x) \,\&\, \neg p_1(x) \to z(x)))$$
$$\&\, bird(tweety) \,\&\, \neg z(tweety) \,\&\, \forall x(p_1(x) \to ab_1(x))$$
$$\&\, \neg\forall x(ab_1(x) \leftrightarrow p_1(x))).$$

From (8), Tweety is clearly an exception for flying, and it satisfies $ab_1$. Here, we substitute $x = tweety$ for $p_1$ corresponding to $ab_1$. Thus, from $Circum(A_2; ab_1; fly)$, (10) is derived.

(10) $\forall x(ab_1(x) \leftrightarrow (x = tweety))$

Consequently, the conclusion drawn from (7) is withdrawn. And substituting (10) for (1) yields:

(11) $\forall x(bird(x) \,\&\, \neg(x = tweety) \to fly(x))$

which says that if $x$ is a bird and is not Tweety then it flies. This is the formalization of non-monotonic reasoning, which can be described in classical logic.

After McCarthy's formula circumscription, several variants have been developed. For example, Lifschitz proposed *prioritized circumscription* capable of giving a priority of minimization of predicates in [20]. Such a mechanism is of special importance for the formalization of common-sense reasoning.

Indeed many non-monotonic logics have been developed from different perspectives, it is necessary to seek general properties of non-monotonic reasoning. This is to work out a meta-theory for non-monotonic reasoning. If meta-theory is established, we can understand the relations of different non-monotonic logics and it is appropriate for applications.

One of the fundamental approaches to reasoning (or logic) is to study consequence relation. It can be seen as axiomatizing consequence relations. Gabbay firstly investigated a meta-theory for non-monotonic reasoning in this direction in Gabbay [21].

He axiomatized *non-monotonic inference relation* based on the minimal conditions it should satisfy. In other words, non-monotonic inference relation $\mid\!\sim$ is an inference relation satisfying the following three properties:

(1) $A \in \Gamma \Rightarrow \Gamma \mid\!\sim A$
(2) $\Gamma \mid\!\sim A$ and $\Gamma \mid\!\sim B \Rightarrow \Gamma \mid\!\sim A\&B$
(3) $\Gamma \mid\!\sim A$ and $\Gamma, A \mid\!\sim B \Rightarrow \Gamma \mid\!\sim B$

Here $\Gamma$ is a set of formulas and $A$ and $B$ are a formula. (1) is called *Reflexivity*, (2) is called *Restricted Monotonicity*, and (3) is called *Cut*, respectively.

Non-monotonic inference relation $\mid\!\sim$ is related to monotonic inference relation $\vdash$ by (4).

(4) $\Gamma \vdash A \Rightarrow \Gamma \mid\!\sim A$

But the converse of (4) does not hold.

Makinson studied a general theory for non-monotonic inference relations with model theory and established some completeness results; see Makinson [22, 23] for details.

Meta-theory for non-monotonic reasoning can be also advanced semantically. The first such approach was done by Shoham [24]. Shoham developed *preference logic* **P** based on *preference consequence relation*.

**P** can be formalized by the following axioms:

$\vdash A \leftrightarrow B$ and $A \mid\!\sim C \;\Rightarrow\; B \mid\!\sim C$ (Left Logical Equivalence: LLE)
$\vdash A \rightarrow B$ and $C \mid\!\sim A \;\Rightarrow\; C \mid\!\sim B$ (Right Weakning: RW)
$A \mid\!\sim A$ (Reflexivity)
$A \mid\!\sim B$ and $A \mid\!\sim C \;\Rightarrow\; A \mid\!\sim B\&C$ (And)
$A \mid\!\sim C$ and $B \mid\!\sim C \;\Rightarrow\; A \vee B \mid\!\sim C$ (Or)
$A \mid\!\sim B$ and $A \mid\!\sim C \;\Rightarrow\; A\&B \mid\!\sim C$ (Cautious Monotonicity: CM)

Here, $\vdash$ denotes classical consequence relarion.

Later, Kraus, Lehmann and Magidor proved that **P** is complete for preference model in [25]. Additionally, they investigate the system **R** which extends **P** with the axiom (Rational Monotonicity; RM):

$A \mid\!\sim C$ and $A \not\mid\!\sim \neg B \;\Rightarrow\; A\&B \mid\!\sim C$ (Rational Monotonicity: RM).

Preference consequence relation satisfying RM is called *rational consequence relation*. Lehmann and Magidor proposed a theory of non-monotonic consequence relation for knowledge bases in [26].

*Knowledge base* is a set of knowledge with an inference engine, and it is considered as a generalization of *database*. The first formal model of database was proposed by Codd [27], and it is the underlying basis of *relational database*. However, relational database is not suited for knowledge base, since relational database has no inference mechanisms needed for AI systems.

Therefore, a theory of knowledge base was worked out. Logic-based approaches to knowledge base is called *logic database* or *deductive database*. It can precisely model knowledge base in terms of formal logic, and logic programming language like Prolog can serve as a query language.

Logic database is viewed as an extension of relational database. A relation in relational database can be expressed as *tuple*. For example, assume that the tuple (0, 1) is in the relation $R$ in which the first element is less than the second one. Thus, relational data (0, 1) corresponds to the predicate $R(0, 1)$ in first-order predicate logic. It is possible to regard a set of relational data as a set of formulas.

This implies a database can be formalized as a theory in first-order logic, and a query is identified with deduction. The feature has a merit in that both a data model and query can be described in first-order logic.

However, to use logic database as knowledge base, we need to represent negative information and non-monotonic reasoning. Logic database is expressed as a set of formulas, i.e., logic program, but we need more than facts and rules.

To proceed our exposition, we have to give basics of *logic programming*: see Kowalski [28, 29]. In 1974, Kowalski proposed to use predicate logic as a programming language by using Robinson's [30] *resolution principle*. This is a starting point of logic programming. The first logic programming language called *Prolog* was implemented in 1972; see Colmerauer [31].

Logic programming is based on the subset of first-order logic, called Horn clause logic. A *clause* is of the form:

(1) $A_1, ..., A_k \leftarrow B_1, ..., B_n$

Here, $A_1, ..., A_k, B_1, ..., B_n$ are an atomic formula and $\leftarrow$ denotes implication. All variables occurring in $A_1, ..., A_k, B_1, ..., B_n$ are universally quantified. Thus, (1) is equivalent to (2) in classical logic:

(2) $\forall((B_1 \& ... \& B_n) \rightarrow (A_1 \vee ... \vee A_k))$

*Horn clause* is a clause which has at most one positive literal, and is classified as the following three forms. A *program clause* is of the form (3):

(3) $A \leftarrow B_1, ..., B_n$

where $A$ is called *head* and $B_1, ..., B_n$ is called *body*, respectively.

*Unit clause* is of the form (4):

(4) $A \leftarrow$

*Goal clause* is of the form (5):

(4) $\leftarrow B_1, ..., B_n$

A *logic program* is a finite set of Horn clauses. Computation of logic programming is performed by resolution principle, i.e., SLD-resolution.

There are two types of semantics for logic programs, i.e., *declarative semantics*, and *operational semantics*.

Declarative semantics declaratively interprets the meaning of programs in terms of mathematical structures. We have *model-theoretic semantics* and *fixpoint semantics* as declarative semantics. Operational semantics describes the input-output relation of programs, which corresponds to the interpreter of programming language.

Van Emden and Kowalskip [32] systematically studied these semantics for Horn clause logic programs and proved these equivalence. Let $S$ be a set of closed formulas of Horn logic. Then, *Herbrand model* for $S$ is *Herbrand interpretation* for $S$. A model-theoretic semantics for logic programs is given by means of the *least Herbrand model*.

We here write $M(A)$ for all sets of Herbrand models for $A$. Then, the intersection $\cap M(A)$ of all Herbrand models for $A$ is also Herbrand model. This is called the *model intersection property*.

The intersection of all Herbrand models for the program $P$ is called the least Herbrand model, denoted $M_P$. Van Emden and Kowalski proved the following result:

**Theorem 5.3** *Let $P$ be a logic program and $B_P$ be Herbrand base of $P$. Then, the following holds:*

$$M_P = \{A \in B_P \mid A \textit{ is a logical consequence of } P\}.$$

From Theorem 5.3, we can see that atomic formulas belonging to $B_P$ is a logical consequence of $P$. Note that Theorem 5.3 does not in general hold for non-Horn clause.

Fixpoint semantics defines the meaning of recursive procedure by the least fixpoint. Since the computation rules of logic programming are recursively applied to goal clauses, it is possible to give fixpoint semantics. Observe that fixpoint semantics is a basis of *denotational semantics* for programming languages; see Stoy [33] for details.

In fixpoint semantics, the concept of *lattice* plays an important role. Let $S$ be a set and $R$ is a binary relation on $S$. $(S, R)$ is a lattice if for any $s, t \in S$ there are their *least upper bound* (lub) denoted $s \vee t$ and *greast lower boud* (glb) denoted $s \wedge t$. $S$ is a *complete lattice* if $S$ is an ordered relation and for each subset $X \subseteq S$ there exist both lub and glb.

Now, we define a mapping $T : S \rightarrow S$ for complete lattie $(S, \leq)$. We say that a mapping $T$ is *monotonic* when $T(s) \leq T(s')$ if $s \leq s'$. We say that $T$ is *continuous* if $T(lub(X)) = lub(T(X))$ for every directed subset $X \subseteq S$.

We say that $s \in S$ is a *fixpoint* of $T$ if $T(s) = s$. If $s \in S$ is a fixpoint of $T$ and for every $s' \in S$ if $T(s') = s'$ then $s \leq s'$, then $T$ is the *least fixpoint* $lfp(T)$ of $T$. If $s \in S$ is a fixpoint of $T$ and for every $s' \in S$ if $T(s') = s'$ then $s \geq s'$, then $T$ is the *greatest fixpoint* $gfp(T)$ of $T$.

Knaster-Tarski's *fixpoint theorem* states there exists a fixpont of monotonic mapping on complete lattice. We here define transfinite sequences $T \uparrow \omega$ and $T \downarrow \omega$ for lattice $(S, \leq)$ as follows:

$T \uparrow 0 = T(\bot)$
$T \uparrow n+1 = T(T \uparrow n) \ \ (0 \leq n)$
$T \uparrow \omega = lub(\{T \uparrow n \mid 0 \leq n\})$
$T \downarrow 0 = T(\top)$
$T \downarrow n+1 = T(T \downarrow n) \ \ (0 \leq n)$
$T \downarrow \omega = glb(\{T \downarrow n \mid 0 \leq n\})$

where $\bot$ is the minimal element of $S$ and $\top$ is the maximal element, respectively.

If $T$ is monotonic and continuous, then we have the following:

$lfp(T) = T \uparrow \omega$
$gfp(T) = T \downarrow \omega$

A basic idea of fixpoint semantics is to interpret the meaning of recursive procedures by fixpoint. Let $P$ be a program and $2^{B_P}$ be a set of its all Herbrand interpretations. Then, the set is complete lattice with $\subseteq$.

Next, we define transformation function $T_P : B_P \rightarrow B_P$ corresponding to the operation of procedures. Let $I$ be an interpretation and $P$ be a program. Then, $T_P$ is defined as follows:

$T_P(I) = \{A \in B_P \mid A \leftarrow A_1, ..., A_n \ is \ a \ ground \ instance \ in \ P \ and$
$\{A_1, ..., A_n\} \subseteq I\}$

The intuitive meaning of $T_P$ is the operator on Herbrand base which corresponds to the resolution rule.

In practice, an interpreter for logic programming language computes an answer by applying the rule corresponding to $T_P$ in a finite time.

**Theorem 5.4** $M_P = lfp(T_P) = T_P \uparrow \omega$.

From Theorems 5.3 and 5.4, the equivalence of model-theoretic and fixpoint semantics is established.

**Theorem 5.5** *The Herbrand interpretation I of a set A of procedures in a logic program is a model of A iff I is closed with respect to the transformation T associated with A.*

Operational semantics for logic programming is described by resolution principle. The equivalence of operational and model-theoretic semantics follows from the completeness of resolution.

The resolution principle for logic programming is called the *SLD resolution*. It uses *SL resolution* for definite (Horn) clauses.

As a result, van Emden and Kowalski proved the equivalence of above three semantics for Horn clause logic programming; see [32]. Later, Apt and van Emden extended their results and proved soundness and completeness of SLD resolution using fixpoint semantics in [34].

Let $P$ be a program, $G$ be a goal and $R$ be a computation rule. A *SLD derivation* is a finite (or infinite) sequence of goals $G_1, G_2, ...$, sequences of variants of goals $C_1, C_2, ...$, and sequences of most general unifier (mgu) $\theta_1, \theta_2, ...$ . Here, the variant $C_{i+1}$ is derivable from $G_i$ and $C_i$ using $\theta_i$ and $R$. Computation rule $R$ specifies how to select a literal from a goal.

An *SLD refutation* is a finite derivation of $P \cup \{G\}$ with $R$ in which the last goal is the empty clause. A *success set* of program $P$ is a set of $A \in B_P$ in which there exists a SLD refutation of $P \cup \{\leftarrow A\}$.

Clark [35] proved soundness of SLD resolution.

**Theorem 5.6** *If there exists a SLD derivation of* $P \cup \{G\}$ *using R, then* $P \cup \{G\}$ *is unsatisfiable.*

Consequently, the success set of a program is included in the least Herbrand model. Theorem 5.7 states soundness of SLD resolution.

**Theorem 5.7** *If* $P \cup \{G\}$ *is unsatisfiable, then there exists a SLD derivation of* $P \cup \{G\}$ *with R.*

In logic programming, negation is introduced by means of *negation as failure* (NAF) as introduced by Clark [35]. Note that NAF is different from classical negation. In Prolog, NAF is defined as follows:

```
not(P) <- P,  !,  fail
not(P) <-
```

Here, the procedure `fail` guarantees failure. In this definition, the first clause proves `P` and fails by `fail`. If the first clause fails, then by the second clause, `not(P)` is proved. The second clause is exactly the definition of CWA.

Clark proposed to interpret NAF within classical logic. Since in NAF $negA$ is proved when $A$ is finitely fails, then we need to formalize the concept of "finite failure".

A finitely failed SLD tree of $P \cup \{G\}$ is a finite tree which contains success leaves. Thus, *SLD finitely failed set* is a set of $A \in B_P$ in which there exists a finitely failed tree of $P \cup \{G\}$,

A SLD finitely failed set $F_P$ is defined as follows;

$F_P = B_P \setminus T_P \downarrow \omega$.

SLD derivation is *fair* if every literal in the derivation is selected in finite step.

**Theorem 5.8** *Let $P$ be a program and $A \in B_P$. Then, the following are equivalent:*

*(1)* $A \in F_P$
*(2)* $A \notin T_P \downarrow \omega$
*(3) SLD tree for $P \cup \{\leftarrow A\}$ with every fair computation rule $R$ is finitely fails.*

Clark introduced the notion of *completion* to give an interpretation of NAF in classical logic. Completion provides the if-and-only-if definition of a clause. Since a clause uses the if definition, it cannot derive negative information.

To deal with *complete database*, Clark extended the form of a clause as follows:

$$A(t_1, ..., t_n) \leftarrow L_1, ..., L_m$$

Then, we can transform a clause into the following general form:

$$A(t_1, ..., t_n) \leftarrow \exists y_1 ... \exists y_k (x_1 = t_1 \& .... \& x_n = t_n \& L_1 \& ... \& L_m)$$

Here, $x_1, ..., x_n$ denote new variables and $y_1, ..., y_k$ variables in the orginal clause, respectively.

If the original form of a program is:

$$A(x_1, ..., x_n) \leftarrow E_1$$

...

$$A(x_1, ..., x_n) \leftarrow E_j$$

then, *complete definition* of a predicate $A$ is expressed as follows:

$$A(x_1, ..., x_n) \leftrightarrow E_1 \vee ... \vee E_j$$

The complete database $comp(P)$ of a program $P$ consists of a set of complete definition of all predicates in $P$ and assumes the following axioms:

(1) For all different constants $c, d, c \neq d$
(2) For all different function symbols $f, g, f(x_1, ..., x_n) \neq g(x_1, ..., x_n)$
(3) For every constant $c$ and function symbol $f, f(x_1, ..., x_n) \neq c$
(4) For every term $t(x)$ containing $x, t(x) \neq x$

(5) For every function symbol $f$, $(x_1 \neq y_1 \vee ... \vee x_n \neq y_n) \rightarrow f(x_1, ..., x_n) \neq f(y_1, ..., y_n)$
(6) $x = x$
(7) For every function symbol $f$, $(x_1 = t_1 \& ... \& x_n = t_n) \rightarrow f(x_1, ..., x_n) = f(t_1, ..., t_n)$
(8) For every predicate symbol $A$, $(x_1 = t_1 \& ... \& x_n = t_n) \rightarrow A(x_1, ..., x_n) \rightarrow A(t_1, ..., t_n)$.

Clearly, $P$ is a logical consequence of $comp(P)$. The following is an example of completion. Let $P$ be a program given by:

```
mortal(X) <- man(X)
man(socrates) <-
man(aristotle) <-
```

Let `<- mortal(socrates)` be a goal. Then, unifying with the first clause, we have:
`<- man(socrates)`
where mgu is `X = socrates`. Next, the new goal unifies the second clause and the empty clause is derived:
`<-`

Then, the original goal succeeds and we can conclude that Socrates is mortal. Here, we give a complettion of $P$:
$man(X) \leftrightarrow (X = socrates) \vee (X = aristotle)$

From $comp(P)$, it is possible to prove both $man(socrates)$ and $man(aristotle)$. If we consider contraposition of $comp(P)$, we get:
$\neg man(X) \leftrightarrow (X \neq socrates) \& (X \neq aristotle)$

Consequently, we can classically prove $\neg man(mary)$. If $P$ is a Horn clause logic progrm, then $comp(P)$ is conistent, but it is generally not the case for non-Horn clause logic programs.

For complete logic programs, the following theorem holds:

**Theorem 5.9** *Let $P$ be a program and $A \in B_P$. Then, $A \notin gfp(T_P)$ iff there exist no Herbrand models for $comp(P) \cup \{A\}$.*

The operational semantics for NAF is called *SLDNF resolution* or *query evaluation procedure*, which extends SLD resolution with NAF. In SLDNF resolution, when the computation rule $R$ selects a negative literal, it always ground one to avoid infinite loop, i.e., *floundering*. The computation rule satisfying this condition is called *safe*.

An *SLDNF derivation* of $P \cup \{G\}$ with $R$ is a finite sequence of goals $G_1 = G, G_2, ...$, a finite sequence of variants of program clause $C_1, C_2, ...$, and a finite sequence of mgu $\theta_1, \theta_2, ...$ .

Here, $G_{i+1}$ is derivable from $G_i$, and there are two possibilities.

(1) $G_1$ is $\leftarrow L_1, ..., L_m$, $R$ selects a positive literal $L_k$. Aussume $A \leftarrow B_1, ..., B_n$ is unifiable input clause by $L_k$ and $A$ with mgu $\theta$. Then, $G_{i+1}$ is $\leftarrow (L_1, ...., L_{k-1}, B_1, ..., B_n, L_{k+1}, ..., L_m)\theta$.
(2) $G_i$ is $\leftarrow L_1, ..., L_m$, $R$ selects negative literal $L_k = \neg A$. If $\leftarrow A$ succeeds, then $\neg A$ fails and the goal $G_i$ also fails. If $\leftarrow \neg A$ succeeds (i.e., $\leftarrow A$ finitely fails), then $G_{i+1}$ is the one from $G_i$ by deleting $L_k$,
i.e., $\leftarrow L_1, ..., L_{k+1}, L_{k+1}, ..., L_m$.

An *SLDNF refutation* of $P \cup \{G\}$ with $R$ is a finite SLDNF derivation of $P \cup \{G\}$ with $R$ and its last goal is the empty clause.

Clark proved soundness of SLDNF resolution:

**Theorem 5.10** *Let $P$ be a general program, $G$ be a general goal, and $R$ is safe computation rule. If $P \cup \{G\}$ fails, then $G$ is a logical consequence of $comp(P)$.*

The converse of Theorem 5.10, i.e., completeness, was proved by Jaffar, Lassez and Lloyd [36], provided that the computation rules are restricted to fair ones.

**Theorem 5.11** *If $A \in F_P$, then $\neg A$ is a logical consequence of $comp(P)$.*

In logic programming, we have three means to derive negative information, i.e., NAF, CWA, and the Herbrand rule (cf. Lloyd [37, 38]). We have already discussed NAF and CWA. *Herbrand rule* was introduced by Lloyd, which assumes $\neg A$ if there exist no Herbrand models for $comp(P) \cup \{A\}$.

Lloyd summarized the relationships of these three rules.

**Theorem 5.12** $\{A \in B_P \mid \neg A \ is \ derived \ by \ \text{NAF}\} = B_P \setminus T_P \downarrow \omega$.
$\{A \in B_P \mid \neg A \ is \ derived \ by \ Herbrand \ rule\} = B_P \setminus gfp(T_P)$.
$\{A \in B_P \mid \neg A \ is \ derived \ by \ \text{CWA}\} = B_P \setminus T_P \uparrow \omega$.

From Theorem 5.12, we can understand that CWA is the strongest rule, and the next is CWA and the weakest is NAF. Logic programming can in fact handle negative information. Recently, further extensions of logic programming have been done, but we omit their survey here.

Logic database is based on first-order logic which can be used both as data model and query language. And it is implemented by logic programming language. However, if we consider logic database as knowledge base then it has to represent negative information and non-monotonic reasoning.

As in relational database, logic database needs *integrity constraint* (IC), which is described by goal clause. IC describes the properties a database should satisfy. Logic database $DB$ satisfies IC when IC is a logical consequence of $comp(DB)$. Thus, we can see that $DB$ must incorporate NAF. Reiter [39] have proposed a formal model for logic database,

There are two approaches to logic database, namely proof-theoretic approach and Model-theoretic approach. The proof-theoretic approach corresponds to the operational semantics for logic programming, we need the following three meta-rules to enhance expressive power:

- *closed world assumption* (CWA)
- *unique name axiom* (UNA)
- *domain closure axiom* (DCA)

CWA is a default rule to derive negative information, as discussed above.

(CWA) $\nvdash A \Rightarrow \vdash \neg A$

where $A$ is a ground literal. It means that the information not derived from a database is assumed to be false by interpreting that the information in a database is ideally represented.

UNA assumes different entities have different names.

(UNA) $a_1 \neq a_2$ & $a_1 \neq a_3$ & $a_1 \neq a_4$ &...

DCA assumes that existing entities belong to the domain.

(DCA) $\forall x(x = a_1 \vee x = a_2 \vee ...)$

In addition to these meta-rules, we assume the axioms of completion.

The model-theoretic approach to logic database is based on the model-theoretic semantics for logic programming. Therefore, we can formalize it by means of the least Herbrand model.

By CWA, the positive literal which is false in the least Herbrand model is interpreted as its negation. UNA assumes that in Herbrand interpretations constants in the language agree with constants in the domain. By DCA, we mean that constants which are not in the language do not exist in the domain in Herbrand interpretations. It is also obvious that the least Herbrand model of DB is also the one of IC.

We denote by $M$ the least Herbrand model of logic database $DB$. Then, a query $A(x)$ represents a set of constants $c$ such that $A(c)$ is true in $M$. Let $\|A(x)\|$ be the truth-value of a query $A(x)$. Then, we have the following:

$\|A(x)\|; = \{c \mid \models_M A(c)\}$.

Thus, the truth-value of CWA is:

$\|\neg A(x)\| = \{c \mid \not\models_M A(c)\}$

which means that positive ground literal $A(c)$ is false in $M$.

Here, we focus on non-monotonic reasoning in logic database. CWA is not classical negation, and it is in fact a non-monotonic rule.

Consider the following database $DB$:

$DB = \{A \rightarrow B\}$

Neither $A$ nor $B$ can be derived from $DB$, namely

(1) $DB \nvdash A$
(2) $DB \nvdash B$

Then, by CWA, (3) and (4) are derived:

(3) $DB + CWA \vdash \neg A$

(4) $DB + CWA \vdash \neg B$

Next, the new information $A$ is added.

(5) $DB' = DB \cup \{A\} = \{A, A \rightarrow B\}$

From (5), (6) holds.

(6) $DB' + CWA \vdash B$

Thus, $\neg B$ derived in (4) is withdrawn. CWA is a simple non-monotonic rule. This means that logic database can serve as knowledge base.

Unfortunately, CWA does not hold for non-Horn database. Consider the following database $DB$: $DB = \{A \vee B\}$ By CWA, the following hold:

(1) $DB + CWA \vdash \neg A$
(2) $DB + CWA \vdash \neg B$

From (1) and (2), by CWA, we have the following:

(3) $DB + CWA \vdash \neg A \& \neg B$

(3) is equivalent to (4).
(4) $DB + CWA \vdash \neg(A \vee B)$

But, (4) is inconsistent with $DB$. As in the example, CWA cannot adequately handle disjunctive information. This implies that logic data base should be extended to overcome the defect.

*Indefinite deductive database*, as proposed by Minker [40], improves deductive database. Data in the form of non-Horn clause are allowed and CWA is replaced by *generalized closed world assumption* (GCWA).

GCWA is of the form:

(GCWA) $\nvdash A \vee C \;\Rightarrow \vdash \neg A$

where $C$ is included in the set $E$ of positive clauses which are not provable and $\nvdash C$.

Consider the following $IDDB$:

$IDDB = \{A \vee B \vee \neg C\}$

By GCWA, (1) is derived:

(1) $IDDB + GWCA \vdash \neg C$

Assume that the information $C$ is added. Then, (2) holds.

(2) $IDDB \cup \{C\} \nvdash \neg C$.

which means GCWA is, like CWA, a non-monotonic rule.

GWCA can also solve the problem of *null value*. Its formalization considered to be very important in incomplete database. Consider the following $DB$:

$DB = \{A(\omega), B(a), B(b)\}$

where $\omega$ denotes a null-value (or skolem constant). Now, suppose the domain $D = \{a, b\}$. Then, the existential statement $\exists x A(x)$ is expressed as (1):

(1) $\exists x A(x) \leftrightarrow A(a) \vee A(b)$

Here, it can be considered that $A(\omega)$ is shorthand for $A(a) \vee A(b)$. Thus, if we give a null value $A(a) \vee A(b)$, then from (1), we can employ GCWA to derive negation. Namely, we cannot deduce $\neg A(a)$ from (1), and it is compatible to our intuition.

Another approach for disjunctive information is the *stratified database* due to Apt, Blair and Walker [41]. It is based on the idea that we can logically interpret negation in deductive database when NAF and classical negation agree.

In stratified database, the representation of disjunctive information leads different derivations of negation. Consider the following $IDDB$:

$IDDB = \{A \vee B\}$

which can be written in classical logic in two forms:

(1) $IDDB_1 = \{A \leftarrow \neg B\}$
(2) $IDDB_2 = \{B \leftarrow \neg A\}$

In (1), the stratification is $(\{B\}, \{A\})$, and we denote by $DB_1$ the stratified database. Then, (3) holds:

(3) $DB_1 \vdash A \& \neg B$

In (2), the stratification is $(\{A\}, \{B\})$ and we denote by $DB_2$ the stratified database. Then, (4) holds:

(4) $DB_2 \vdash B \& \neg A$

There are intimate connections between stratified database and non-monotonic reasoning, but we omit the discussion here.

The semantics of logic programs in connection with non-monotonic reasoning is hot area. The *stable model semantics* was proposed by Gelfond and Lifschitz [42] for NAF. Van Gelder et al. [43] proposed *well-founded semantics* for NAF.

Gelfond and Lifschitz proposed *extended logic programming*, which has two kinds of negation, i.e., NAF and classical (explicit) negation in [44] and its semantics called the *answer set semantics*. A similar extension may be in *logic programming with exceptions* of Kowalski and Sadri [45]. Unfortunately, logically speaking, their explicit negation in both extensions is not classical negation.

We gave an exposition of the existing non-monotonic logics. But, none of them can be regarded as a logical system in the sense of standard logic. We prefer to use formal logic as a basis for common-sense reasoning.

As shown in the previous section, our granularity-based framework can properly represent non-monotonicity based on rough set theory. We can thus claim that one dispenses with non-monotonic logics.

## 5.5 Paraconsistency, Chellas's Conditional Logics, and Association Rules

Paraconsistency and its dual, parcompleteness are now counted as key concepts in intelligent decision systems because so much inconsistent and incomplete information can be found around us.

In this section, a framework of conditional models for conditional logic and their measure-based extensions are introduced in order to represent association rules in a logical way. Then, paracomplete and paraconsistent aspects of conditionals are examined in the framework.

Finally we apply conditionals into the definition of association rules in data mining with confidence and consider their extension to the case of Dempster-Shafer theory of evidence serving double-indexed confidence.

In classical logic, inconsistency means triviality in the sense that all sentences become theorems. Paraconsistency means inconsistency but non-triviality. Thus, we need new kinds of logic like paraconsistent and annotated logics [46]. Paracompleteness is the dual concept of paraconsistency where the excluded middle is not true.

We now put association rules in a framework of conditional models [47] and their measure-based extensions (cf. [1, 2]) and examine their paracomplete and paraconsistent aspects in the framework. Then, we notice that the standard confidence [48] is nothing but a conditional probability where even weights are *a priori* assigned to each transaction that contains the items in question at the same time.

All of such transactions, however, do not necessarily give us such evidence because some co-occurrences might be contingent. For describing such cases we further introduce double-indexed confidence based on Dempster-Shafer theory [49, 50]. Here, we give standard and minimal conditional models.

Given a finite set $\mathscr{P}$ of items as *atomic sentences*, a *language* $\mathscr{L}_{\mathrm{CL}}(\mathscr{P})$ for conditional logic is formed from $\mathscr{P}$ as the set of sentences closed under the usual propositional operators such as $\top, \bot, \neg, \wedge, \vee, \to$, and $\leftrightarrow$ as well as $\Box\!\!\rightarrow$ and $\Diamond\!\!\rightarrow$ [1] (*two conditionals*) in the following usual way.

1. If $x \in \mathscr{P}$ then $x \in \mathscr{L}_{\mathrm{CL}}(\mathscr{P})$.
2. $\top, \bot \in \mathscr{L}_{\mathrm{CL}}(\mathscr{P})$.
3. If $p \in \mathscr{L}_{\mathrm{CL}}(\mathscr{P})$ then $\neg p \in \mathscr{L}_{\mathrm{CL}}(\mathscr{P})$.
4. If $p, q \in \mathscr{L}_{\mathrm{CL}}(\mathscr{P})$ then $p \wedge q$, $p \vee q$, $p \to q$, $p \leftrightarrow q$, $p \Box\!\!\rightarrow q$, $p \Diamond\!\!\rightarrow q \in \mathscr{L}_{\mathrm{CL}}(\mathscr{P})$.

Chellas [47] introduces two kind of models called standard and minimal. Their relationship is similar to Kripke and Scott-Montague models for modal logics.

**Definition 5.3** A *standard conditional model* $\mathscr{M}_{\mathrm{CL}}$ for conditional logic is a structure $\langle W, f, v\rangle$ where $W$ is a non-empty set of possible worlds, $v$ is a truth-assignment function $v : \mathscr{P} \times W \to \{0, 1\}$, and $f$ is a function $f : W \times 2^W \to 2^W$.

[1] In [47], Chellas used only $\Box\!\!\rightarrow$. The latter connective $\Diamond\!\!\rightarrow$ follows Lewis [51].

The truth conditions for $\Box\!\!\rightarrow$ and $\Diamond\!\!\rightarrow$ in standard conditional models are given by

1. $\mathcal{M}_{\mathrm{CL}}, w \models p\Box\!\!\rightarrow q \overset{\mathrm{def}}{\Longleftrightarrow} f(w, \|p\|^{\mathcal{M}_{\mathrm{CL}}}) \subseteq \|q\|^{\mathcal{M}_{\mathrm{CL}}}$,
2. $\mathcal{M}_{\mathrm{CL}}, w \models p\Diamond\!\!\rightarrow q \overset{\mathrm{def}}{\Longleftrightarrow} f(w, \|p\|^{\mathcal{M}_{\mathrm{CL}}}) \cap \|q\|^{\mathcal{M}_{\mathrm{CL}}} \neq \emptyset$,

where $\|p\|^{\mathcal{M}_{\mathrm{CL}}} = \{w \in W \mid \mathcal{M}_{\mathrm{CL}}, w \models p\}$. Thus, we have the following relationship between the two kind conditionals:

$$p\Box\!\!\rightarrow q \leftrightarrow \neg(p\Diamond\!\!\rightarrow\neg q).$$

The function $f$ can be regarded as a kind of selection function. That is, $p\Box\!\!\rightarrow q$ is true at a world $w$ when $q$ is true at any world selected by $f$ with respect to $p$ and $w$. Similarly, $p\Diamond\!\!\rightarrow q$ is true at a world $w$ when $q$ is true at least at one of the worlds selected by $f$ with respect to $p$ and $w$.

A *minimal conditional models* is a Scott-Montague-like extension of standard conditional model [47].

**Definition 5.4** A *minimal conditional model* $\mathcal{M}_{\mathrm{CL}}$ for conditional logic is a structure $\langle W, g, v\rangle$ where $W$ and $v$ are the same ones as in the standard conditional models. The difference is the second term $g : W \times 2^W \to 2^{2^W}$.

The truth conditions for $\Box\!\!\rightarrow$ and $\Diamond\!\!\rightarrow$ in a minimal conditional model are given by

1. $\mathcal{M}_{\mathrm{CL}}, w \models p\Box\!\!\rightarrow q \overset{\mathrm{def}}{\Longleftrightarrow} \|q\|^{\mathcal{M}_{\mathrm{CL}}} \in g(w, \|p\|^{\mathcal{M}_{\mathrm{CL}}})$,
2. $\mathcal{M}_{\mathrm{CL}}, w \models p\Diamond\!\!\rightarrow q \overset{\mathrm{def}}{\Longleftrightarrow} (\|q\|^{\mathcal{M}_{\mathrm{CL}}})^{\mathrm{C}} \notin g(w, \|p\|^{\mathcal{M}_{\mathrm{CL}}})$,

Thus, we have also the following relationship:

$$p\Box\!\!\rightarrow q \leftrightarrow \neg(p\Diamond\!\!\rightarrow\neg q).$$

Note that, if the function $g$ satisfies the following condition

$X \in g(w, \|p\|^{\mathcal{M}_{\mathrm{CL}}}) \Leftrightarrow \cap g(w, \|p\|^{\mathcal{M}_{\mathrm{CL}}}) \subseteq X$

for every world $w$ and every sentence $p$, then, by defining

$f_g(w, \|p\|^{\mathcal{M}_{\mathrm{CL}}}) \overset{\mathrm{def}}{=} \cap g(w, \|p\|^{\mathcal{M}_{\mathrm{CL}}})$,

we have the standard conditional model $\langle W, f_g, V\rangle$ that is equivalent to the original minimal model.

Next, we introduce measure-based extensions of the previous minimal conditional models. Such extensions are models for graded conditional logics.

Given a finite set $\mathcal{P}$ of items as *atomic sentences*, a *language* $\mathcal{L}_{\mathrm{gCL}}(\mathcal{P})$ for graded conditional logic is formed from $\mathcal{P}$ as the set of sentences closed under the usual propositional operators such as $\top, \bot, \neg, \wedge, \vee, \to$, and $\leftrightarrow$ as well as $\Box\!\!\rightarrow_k$ and $\Diamond\!\!\rightarrow_k$ (*graded conditionals*) for $0 < k \le 1$ in the usual way.

1. If $x \in \mathcal{P}$ then $x \in \mathcal{L}_{\mathrm{gCL}}(\mathcal{P})$.
2. $\top, \bot \in \mathcal{L}_{\mathrm{gCL}}(\mathcal{P})$.
3. If $p \in \mathcal{L}_{\mathrm{gCL}}(\mathcal{P})$ then $\neg p \in \mathcal{L}_{\mathrm{gCL}}(\mathcal{P})$.
4. If $p, q \in \mathcal{L}_{\mathrm{gCL}}(\mathcal{P})$ then $p \wedge q, p \vee q, p \to q, p \leftrightarrow q \in \mathcal{L}_{\mathrm{gCL}}(\mathcal{P})$,

5. If $[p, q \in \mathcal{L}_{\mathrm{gCL}}(\mathcal{P})$ and $0 < k \leq 1]$ then $p \,\Box\!\!\rightarrow_k q,\ p \,\Diamond\!\!\rightarrow_k q \in \mathcal{L}_{\mathrm{gCL}}(\mathcal{P})$.

A graded conditional model is defined as a family of minimal conditional model (cf. Chellas [47]):

**Definition 5.5** Given a fuzzy measure
$m : 2^W \times 2^W \rightarrow [0, 1]$,
a *measure-based conditional model* $\mathcal{M}^m_{\mathrm{gCL}}$ for graded conditional logic is a structure

$$\langle W, \{g_k\}_{0<k\leq 1}, v\rangle,$$

where $W$ and $V$ are the same ones as in the standard conditional models. $g_k$ is defined by a fuzzy measure $m$ as

$$g_k(t, X) \stackrel{\mathrm{def}}{=} \{Y \subseteq 2^W \mid m(Y, X) \geq k\}.$$

The model $\mathcal{M}^m_{\mathrm{gCL}}$ is called *finite* because so is $W$. Further, we call the model $\mathcal{M}^m_{\mathrm{gCL}}$ *uniform* since functions $\{g_k\}$ in the model does not depend on any world in $\mathcal{M}^m_{\mathrm{gCL}}$. The truth conditions for $\Box\!\!\rightarrow_k$ and $\Diamond\!\!\rightarrow_k$ in a measure-based conditional model are given by

$$\mathcal{M}^m_{\mathrm{gCL}}, t \models p\Box\!\!\rightarrow_k q \text{ iff } \|q\|^{\mathcal{M}^m_{\mathrm{gCL}}} \in g_k(t, \|p\|^{\mathcal{M}^m_{\mathrm{gCL}}}),$$
$$\mathcal{M}^m_{\mathrm{gCL}}, t \models p\Diamond\!\!\rightarrow_k q \text{ iff } (\|q\|^{\mathcal{M}^m_{\mathrm{gCL}}})^C \notin g_k(t, \|p\|^{\mathcal{M}^m_{\mathrm{gCL}}}).$$

The basic idea of these definitions is the same as in fuzzy measure-based semantics for graded modal logic defined in [1, 2]. When we take $m$ as a conditional probability, the truth conditions of graded conditional becomes

$$\mathcal{M}^{\mathrm{Pr}}_{\mathrm{gCL}}, t \models p\Box\!\!\rightarrow_k q \text{ iff } \Pr(\|q\|^{\mathcal{M}^{\mathrm{Pr}}_{\mathrm{gCL}}} \mid \|p\|^{\mathcal{M}^{\mathrm{Pr}}_{\mathrm{gCL}}}) \geq k.$$

We have several soundness results based on probability-measure-based semantics (cf. [1, 2]) shown in Table 5.2.

As Chellas pointed out in his book [47], conditionals $p\Box\!\!\rightarrow q$ (and also $p\Diamond\!\!\rightarrow q$) is regarded as relative modal sentences like $[p]q$ (and also $\langle p\rangle q$). So we first see paraconsistency and paracompleteness in the usual modal setting for convenience.

Let us define some standard language $\mathcal{L}$ for modal logic with two modal operators $\Box$ and $\Diamond$. In [52], we examined some relationship between modal logics and paraconsistency and paracompleteness.

Let us assume a language $\mathcal{L}$ of modal logic as usual. In terms of modal logic, paracompleteness and paraconsistency have a close relation to the following axiom schemata:

**D** $\Box p \rightarrow \neg\Box\neg p$,
**D**$_{\mathrm{C}}$ $\neg\Box\neg p \rightarrow \Box p$,

because they have their equivalent expressions

$\neg(\Box p \wedge \Box\neg p)$,

$\Box p \vee \Box\neg p$,

respectively. That is, given a system of modal logic $\Sigma$, define the following set of sentences

$T = \{p \in \mathscr{L} \mid \vdash_\Sigma \Box p\}$,

where $\vdash_\Sigma \Box p$ means $\Box p$ is a theorem of $\Sigma$. Then the above two schemata mean that, for any sentence $p$

not($p \in T$ and$\neg p \in T$)

$p \in T$ or$\neg p \in T$

respectively, and obviously the former describes the consistency of $T$ and the latter the completeness of $T$. Thus

- $T$ is inconsistent when $\Sigma$ does not contain **D**.
- $T$ is incomplete when $\Sigma$ does not contain $\mathbf{D}_\mathbf{C}$.

A system $\Sigma$ is regular when it contains the following rule and axiom schemata

$p \leftrightarrow q \Rightarrow \Box p \leftrightarrow \Box q$

$(\Box p \wedge \Box q) \leftrightarrow \Box(p \wedge q)$

Note that any normal system is regular.

In [52], we pointed out the followings. If $\Sigma$ is regular, then we have

**Table 5.2** Soundness results of graded conditionals by probability measures

| $0 < k \leq \frac{1}{2}$ | $\frac{1}{2} < k < 1$ | $k = 1$ | Rules and axiom schemata |
|---|---|---|---|
| ○ | ○ | ○ | **RCEA.** $\frac{p \leftrightarrow q}{(p \Box\!\!\rightarrow_k q) \leftrightarrow (q \Box\!\!\rightarrow_k q)}$ |
| ○ | ○ | ○ | **RCEC.** $\frac{q \leftrightarrow q'}{(p \Box\!\!\rightarrow_k q) \leftrightarrow (p \Box\!\!\rightarrow_k q')}$ |
| ○ | ○ | ○ | **RCM.** $\frac{q \rightarrow q'}{(p \Box\!\!\rightarrow_k q) \rightarrow (p \Box\!\!\rightarrow_k q')}$ |
| | | ○ | **RCR.** $\frac{(q \wedge q') \rightarrow r}{((p \Box\!\!\rightarrow_k q) \wedge (p \Box\!\!\rightarrow_k q')) \rightarrow (p \Box\!\!\rightarrow_k r)}$ |
| ○ | ○ | ○ | **RCN.** $\frac{q}{p \Box\!\!\rightarrow_k q}$ |
| | | ○ | **RCK.** $\frac{(q_1 \wedge \cdots \wedge q_n) \rightarrow q}{((p \Box\!\!\rightarrow_k q_1) \wedge \cdots \wedge (p \Box\!\!\rightarrow_k q_n)) \rightarrow (p \Box\!\!\rightarrow_k q)}$ |
| ○ | ○ | ○ | **CM.** $(p \Box\!\!\rightarrow_k (q \wedge r)) \rightarrow (p \Box\!\!\rightarrow_k q) \wedge (p \Box\!\!\rightarrow_k r)$ |
| | | ○ | **CC.** $(p \Box\!\!\rightarrow_k q) \wedge (p \Box\!\!\rightarrow_k r) \rightarrow (p \Box\!\!\rightarrow_k (q \wedge r))$ |
| | | ○ | **CR.** $(p \Box\!\!\rightarrow_k (q \wedge r)) \leftrightarrow (p \Box\!\!\rightarrow_k q) \wedge (p \Box\!\!\rightarrow_k r)$ |
| ○ | ○ | ○ | **CN.** $p \Box\!\!\rightarrow_k \top$ |
| ○ | ○ | ○ | **CP.** $\neg(p \Box\!\!\rightarrow_k \bot)$ |
| | | ○ | **CK.** $(p \Box\!\!\rightarrow_k (q \rightarrow r)) \rightarrow (p \Box\!\!\rightarrow_k q) \rightarrow (p \Box\!\!\rightarrow_k r)$ |
| | ○ | ○ | **CD.** $\neg((p \Box\!\!\rightarrow_k q) \wedge (p \Box\!\!\rightarrow_k \neg q))$ |
| ○ | | | $\mathbf{CD}_\mathbf{C}$. $(p \Box\!\!\rightarrow_k q) \vee (p \Box\!\!\rightarrow_k \neg q)$ |

(1) $(\Box p \wedge \Box \neg p) \leftrightarrow \Box \neg \top$

where $\bot \leftrightarrow \neg \top$ and $\bot$ is falsity constant, which means inconsistency itself. Thus we have triviality:

$T = \mathscr{L}$.

But if $\Sigma$ is not regular, then we have no longer (1), thus, in general

$T \neq \mathscr{L}$,

which means $T$ is paraconsistent. That is, local inconsistency does not generate triviality as global inconsistency.

Next, we apply the previous idea into conditional logics. In conditional logics, the corresponding axiom schemata

**CD.** $\neg((p \Box\!\!\rightarrow q) \wedge (p \Box\!\!\rightarrow \neg q))$
**CD**$_C$ $(p \Box\!\!\rightarrow q) \vee (p \Box\!\!\rightarrow \neg q)$

Given a system $CL$ of conditional logic, define the following set of conditionals (rules):

$R = \{p \Box\!\!\rightarrow q \in \mathscr{L}_{CD} \mid \vdash_{CL} p \Box\!\!\rightarrow q\}$.

where $\mathscr{L}_{CD}$ is a language for conditional logic and $\vdash_{CL} p \Box\!\!\rightarrow q$ means $p \Box\!\!\rightarrow q$ is a theorem of $CL$. Then the above two schemata mean that, for any sentence $p$

not$(p \Box\!\!\rightarrow q \in R$ and $p \Box\!\!\rightarrow \neg q \in R)$
$p \Box\!\!\rightarrow q \in R$ or $p \Box\!\!\rightarrow \neg q \in R$

respectively, and obviously the former describes the consistency of $R$ and the latter the completeness of $R$. Thus, for the set $R$ of conditionals (rules)

- $R$ is inconsistent when $CL$ does not contain **CD**.
- $R$ is incomplete when $CL$ does not contain **CD**$_C$.

Next, we discuss paraconsistency and paracompleteness in association rules. Let $\mathscr{I}$ be a finite set of *items*. Any subset $X$ in $\mathscr{I}$ is called an *itemset* in $\mathscr{I}$. A database is comprised of *transactions*, which are actually obtained or observed itemsets. Formally, we give the following definition:

**Definition 5.6** A *database* $\mathscr{D}$ *on* $\mathscr{I}$ is defined as $\langle T, V \rangle$, where

1. $T = \{1, 2, \cdots, n\}$ ($n$ is the size of the database),
2. $V : T \to 2^{\mathscr{I}}$.

Thus, for each transaction $i \in T$, $V$ gives its corresponding set of items $V(i) \subseteq \mathscr{I}$. For an itemset $X$, its *degree of support* $s(X)$ is defined by

$$s(X) \stackrel{\text{def}}{=} \frac{|\{t \in T \mid X \subseteq V(t)\}|}{|T|},$$

where $|\cdot|$ is a size of a finite set.

**Definition 5.7** Given a set of items $\mathscr{I}$ and a database $\mathscr{D}$ on $\mathscr{I}$, an *association rule* is an implication of the form $X \Longrightarrow Y$, where $X$ and $Y$ are itemsets in $\mathscr{I}$ with $X \cap Y = \emptyset$.

The following two indices were introduced in [48].

**Definition 5.8** 1. An association rule $r = (X \Longrightarrow Y)$ holds with *confidence* $c(r)$ $(0 \leq c(r) \leq 1)$ in $\mathscr{D}$ if and only if

$$c(r) = \frac{s(X \cup Y)}{s(X)}.$$

2. An association rule $r = (X \Longrightarrow Y)$ has a *degree of support* $s(r)$ $(0 \leq s(r) \leq 1)$ in $\mathscr{D}$ if and only if

$$s(r) = s(X \cup Y).$$

Here, we will deal with the former index.

Mining of association rules is actually performed by generating all rules that have certain minimum support (denoted *minsup*) and minimum confidence, denoted *minconf*, that a user specifies. See, e.g., [48] for details of such algorithms for finding association rules.

For example, consider the movie database in Table 5.3, where AH and HM means Ms. Audrey Hepburn and Mr. Henry Mancini, respectively.

If you have watched several (famous) Ms. Hepburn's movies, you might hear some wonderful music composed by Mr. Mancini. This can be represented by the association rule

$$r = \{\text{AH}\} \Longrightarrow \{\text{HM}\}$$

with its confidence

$$c(r) = \frac{s(\{\text{AH}\} \cup \{\text{HM}\})}{s(\{\text{AH}\})} = 0.5$$

and its degree of support

$$s(r) = \frac{|\{T \mid \{\text{AH}\} \cup \{\text{HM}\} \subseteq T\}|}{|\mathscr{D}|} = \frac{4}{100} = 0.04.$$

We now describe measure-based conditional models for databases. Let us regards a finite set $\mathscr{I}$ of items as *atomic sentences*. Then, a *language* $\mathscr{L}_{\text{gCL}}(\mathscr{I})$ for graded conditional logic is formed from $\mathscr{I}$ as the set of sentences closed under the usual propositional operators such as $\top, \bot, \neg, \wedge, \vee, \rightarrow$, and $\leftrightarrow$ as well as $\Box\!\!\rightarrow_k$ and $\Diamond\!\!\rightarrow_k$ (*graded conditionals*) for $0 < k \leq 1$ in the usual way.

1. If $x \in \mathscr{I}$ then $x \in \mathscr{L}_{\text{gCL}}(\mathscr{I})$.
2. $\top, \bot \in \mathscr{L}_{\text{gCL}}(\mathscr{I})$.

**Table 5.3** Movie database

| No. | Transaction (movie) | AH | HM |
|---|---|---|---|
| 1 | Secret people | 1 | |
| 2 | Monte Carlo baby | 1 | |
| 3 | Roman holiday | 1 | |
| 4 | My fair lady | 1 | |
| 5 | Breakfast at Tiffany's | 1 | 1 |
| 6 | Charade | 1 | 1 |
| 7 | Two for the road | 1 | 1 |
| 8 | Wait until dark | 1 | 1 |
| 9 | Days of wine and rose | | 1 |
| 10 | The great race | | 1 |
| 11 | The pink panther | | 1 |
| 12 | Sunflower | | 1 |
| 13 | Some like it hot | | |
| 14 | 12 Angry men | | |
| 15 | The apartment | | |
| | …… | | |
| 100 | Les aventuriers | | |

3. If $p \in \mathscr{L}_{\mathrm{gCL}}(\mathscr{I})$ then $\neg p \in \mathscr{L}_{\mathrm{gCL}}(\mathscr{I})$.
4. If $p, q \in \mathscr{L}_{\mathrm{gCL}}(\mathscr{I})$ then $p \wedge q$, $p \vee q$,$p \rightarrow q$, $p \leftrightarrow q \in \mathscr{L}_{\mathrm{gCL}}(\mathscr{I})$,
5. If $[p, q \in \mathscr{L}_{\mathrm{gCL}}(\mathscr{I})$ and $0 < k \leq 1]$ then $p \,\Box\!\!\rightarrow_k q$, $p \,\Diamond\!\!\rightarrow_k q \in \mathscr{L}_{\mathrm{gCL}}(\mathscr{I})$.

A measure-based conditional model is defined as a family of minimal conditional model (cf. Chellas [47]):

**Definition 5.9** Given a database $\mathscr{D} = \langle T, V \rangle$ on $\mathscr{I}$ and a fuzzy measure $m$, a *measure-based conditional model* $\mathscr{M}^m_{\mathrm{g}\mathscr{D}}$ for $\mathscr{D}$ is a structure $\langle W, \{g_k\}_{0<k\leq 1}, v \rangle$, where (1) $W = T$, (2) for any world (transaction) $t$ in $W$ and any set of itemsets $X$ in $2^{\mathscr{I}}$, $g_k$ is defined by a fuzzy measure $m$ as $g_k(t, X) \stackrel{\mathrm{def}}{=} \{Y \subseteq 2^W \mid m(Y, X) \geq k\}$, and (3) for any item $x$ in $\mathscr{I}$, $v(x, t) = 1$ iff $x \in V(t)$.

The model $\mathscr{M}^m_{\mathrm{g}\mathscr{D}}$ is called *finite* because so is $W$. Further, we call the model $\mathscr{M}^m_{\mathrm{g}\mathscr{D}}$ *uniform* since functions $\{g_k\}$ in the model does not depend on any world in $\mathscr{M}^m_{\mathrm{g}\mathscr{D}}$.

The truth condition for $\Box\!\!\rightarrow_k$ in a grade conditional model is given by

$$\mathscr{M}^m_{\mathrm{g}\mathscr{D}}, t \models p \Box\!\!\rightarrow_k q \text{ iff } \|q\|^{\mathscr{M}^m_{\mathrm{g}\mathscr{D}}} \in g_k(t, \|p\|^{\mathscr{M}^m_{\mathrm{g}\mathscr{D}}}),$$

where

$$\|p\|^{\mathscr{M}^m_{\mathrm{g}\mathscr{D}}} \stackrel{\mathrm{def}}{=} \{t \in W(= T) \mid \mathscr{M}^m_{\mathrm{g}\mathscr{D}}, t \models p\}.$$

The basic idea of this definition is the same as in fuzzy-measure-based semantics for graded modal logic defined in [1, 2].

For example, the usual degree of confidence [48] is nothing but the well-known *conditional probability*, so we define function $g_k$ by conditional probability.

**Definition 5.10** For a given database $\mathscr{D} = \langle T, V \rangle$ on $\mathscr{I}$ and a conditional probability

$$\mathrm{pr}(B|A) = \frac{|A \cap B|}{|A|},$$

its corresponding *probability-based graded conditional model* $\mathscr{M}^{\mathrm{pr}}_{\mathrm{g}\mathscr{D}}$ is defined as a structure $\langle W, \{g_k\}_{0<k\leq 1}, v \rangle$, where $g_k(w, X) \stackrel{\mathrm{def}}{=} \{Y \subseteq 2^W \mid \mathrm{pr}(t(Y) \mid t(X)) \geq k\}$, where $t(X) \stackrel{\mathrm{def}}{=} \{w \in W \mid X \subseteq w\}$.

The truth condition of graded conditional is given by

$$\mathscr{M}^{\mathrm{pr}}_{\mathrm{g}\mathscr{D}}, t \models p \Box\!\!\rightarrow_k q \text{ iff } \mathrm{pr}(\|q\|^{\mathscr{M}^{\mathrm{pr}}_{\mathrm{g}\mathscr{D}}} \mid \|p\|^{\mathscr{M}^{\mathrm{pr}}_{\mathrm{g}\mathscr{D}}}) \geq k.$$

Then, we can have the following theorem:

**Theorem 5.13** *Given a database $\mathscr{D}$ on $\mathscr{I}$ and its corresponding probability-based graded conditional model*

$\mathscr{M}^{\mathrm{pr}}_{\mathrm{g}\mathscr{D}}$, *for an association rule $X \Longrightarrow Y$,*

*we have*

$$c(X \Longrightarrow Y) \geq k \text{ iff } \mathscr{M}^{\mathrm{pr}}_{\mathrm{g}\mathscr{D}} \models p_X \Box\!\!\rightarrow_k p_Y, \text{ where } \mid p_X \mid = X \text{ and } \mid p_Y \mid = Y.$$

We formulated association rules as graded conditionals based on probability. Define the following set of rules with confidence $k$:

$$R_k \stackrel{\mathrm{def}}{=} \{p \Box\!\!\rightarrow_k q \in \mathscr{L}_{gCD} \mid \vdash_{gCL} p \Box\!\!\rightarrow_k q\}.$$

A graded conditional $p \Box\!\!\rightarrow_k q$ is also regarded as a relative necessary sentences $[p]_k q$ and the properties of relative modal operator $[\cdot]_k$ are examined in Murai et al. [1, 2] in the following correspondence:

| Confidence $k$ | Systems |
|---|---|
| $0 < k \leq \frac{1}{2}$ | $EMD_CNP$ |
| $\frac{1}{2} < k < 1$ | $EMDNP$ |
| $k = 1$ | $KD$ |

The former two systems are not regular, so $R_k$ may be paraconsistent. The last one is normal so regular.

For $0 < k \leq \frac{1}{2}$, $R_k$ is complete but for some $p$ and $q$, the both rules $p \Box\!\!\rightarrow_k q$ and $p \Box\!\!\rightarrow_k \neg q$ may be generated. This should be avoided.

For $\frac{1}{2} < k < 1$, $R_k$ is consistent but may be paracomplete.

The standard confidence [48] described above is based on the idea that co-occurrences of items in one transaction are evidence for association between the items. Since the definition of confidence is nothing but a conditional probability, even weights are *a priori* assigned to each transaction that contains the items in question at the same time. All of such transactions, however, do not necessarily give us such evidence because some co-occurrences might be contingent.

Thus, we need a framework that can differentiate proper evidence from contingent one and we introduce *Dempster-Shafer theory of evidence* (D-S theory) [49, 50] to describe such a more flexible framework to compute confidence.

There are a variety of ways of formalizing D-S theory and, here we adopt multivalued-mapping-based approach, which was used by Dempster [49]. In the approach, we need two frames, one of which has a probability defined, and a multi-valued mapping between the two frames. Given a database $\mathscr{D} = \langle T, V \rangle$ on $\mathscr{I}$ and an association rule $r = (X \Longrightarrow Y)$ in $\mathscr{D}$, one of frames is the set $T$ of transactions. Another one is defined by

$$R = \{r, \overline{r}\},$$

where $\overline{r}$ denotes the negation of $r$.

The remaining tasks are (1) to define a probability distribution pr on $T$: $\mathrm{pr} : T \to [0, 1]$, and (2) to define a multi-valued mapping $\Gamma : T \to 2^R$. Given pr and $\Gamma$, we can define the well-known two kinds of functions in Dempster-Shafer theory: for $X \subseteq 2^R$,

$$\mathrm{Bel}(X) \stackrel{\mathrm{def}}{=} \mathrm{pr}(\{t \in T \mid \Gamma(t) \subseteq X\}),$$

$$\mathrm{Pl}(X) \stackrel{\mathrm{def}}{=} \mathrm{pr}(\{t \in T \mid \Gamma(t) \cap X \neq \emptyset\}),$$

which are called *belief* and *plausibility* functions, respectively. Now, we have the following double-indexed confidence:

$$c(r) = \langle \mathrm{Bel}(r), \mathrm{Pl}(r) \rangle.$$

Next, we introduce multi-graded conditional models for databases. Given a finite set $\mathscr{I}$ of items as *atomic sentences*, a *language* $\mathscr{L}_{\mathrm{mgCL}}(\mathscr{I})$ for graded conditional logic is formed from $\mathscr{I}$ as the set of sentences closed under the usual propositional operators as well as $\Box\!\!\rightarrow_k$ and $\Diamond\!\!\rightarrow_k$ (*graded conditionals*) for $0 < k \leq 1$ in the usual way. Note that, in particular,

$$p, q \in \mathscr{L}_{\mathrm{mgCL}}(\mathscr{I}) \text{ and } 0 < k \leq 1) \Rightarrow p\Box\!\!\rightarrow_k q,\ p\Diamond\!\!\rightarrow_k q \in \mathscr{L}_{\mathrm{mgCL}}(\mathscr{I}).$$

**Definition 5.11** Given a database $\mathscr{D}$ on $\mathscr{I}$, a *multi-graded conditional model* $\mathscr{M}_{\mathrm{mg}\mathscr{D}}$ for $\mathscr{D}$ is a structure $\langle W, \{\{\underline{g}_k, \overline{g}_k\}\}_{0<k\leq 1}, v\rangle$, where (1) $W = T$, (2) for any world (transaction) $t$ in $W$ and any set of itemsets $\mathscr{X}$ in $2^{\mathscr{I}}$, $g_k$ is defined by belief and plausibility functions:

**Table 5.4** Soundness results of graded conditionals by belief and plausibility functions

| Belied function | | | Rules and axiom schemata | Plausibility function | | |
|---|---|---|---|---|---|---|
| $0 < k \leq \frac{1}{2}$ | $\frac{1}{2} < k < 1$ | $k = 1$ | | $0 < k \leq \frac{1}{2}$ | $\frac{1}{2} < k < 1$ | $k = 1$ |
| ○ | ○ | ○ | **RCEA** | ○ | ○ | ○ |
| ○ | ○ | ○ | **RCEC** | ○ | ○ | ○ |
| ○ | ○ | ○ | **RCM** | ○ | ○ | ○ |
| ○ | ○ | ○ | **RCR** | | | ○ |
| ○ | ○ | ○ | **RCN** | ○ | ○ | ○ |
| | | ○ | **RCK** | | | |
| ○ | ○ | ○ | **CM** | ○ | ○ | ○ |
| ○ | ○ | ○ | **CC** | | | ○ |
| ○ | ○ | ○ | **CR** | | | ○ |
| ○ | ○ | ○ | **CN** | ○ | ○ | ○ |
| ○ | ○ | ○ | **CP** | ○ | ○ | ○ |
| | | ○ | **CK** | | | |
| | ○ | ○ | **CD** | | | |
| | | | **CD$_C$** | ○ | | |

$$\underline{g}_k(t, \mathscr{X}) \stackrel{\text{def}}{=} \{\mathscr{Y} \subseteq 2^W \mid \mathrm{Bel}(\mathscr{Y}, \mathscr{X}) \geq k\},$$
$$\overline{g}_k(t, \mathscr{X}) \stackrel{\text{def}}{=} \{\mathscr{Y} \subseteq 2^W \mid \mathrm{Pl}(\mathscr{Y}, \mathscr{X}) \geq k\},$$

and (3) for any item $x$ in $\mathscr{I}$, $v(x, t) = 1$ iff $x \in V(t)$.

The model $\mathscr{M}_{\mathrm{mg}\mathscr{D}}$ is called *finite* because so is $W$. Here, we call the model $\mathscr{M}_{\mathrm{mg}\mathscr{D}}$ *uniform* since functions $\{\underline{g}_k, \overline{g}_k\}$ in the model does not depend on any world in $\mathscr{M}_{\mathrm{mg}\mathscr{D}}$.

The truth conditions for $\Box\!\!\rightarrow_k$ and $\Diamond\!\!\rightarrow_k$ are given by

$$\mathscr{M}_{\mathrm{mg}\mathscr{D}}, w \models p\Box\!\!\rightarrow_k q \text{ iff } \|q\|^{\mathscr{M}_{\mathrm{mg}\mathscr{D}}} \in \underline{g}_k(t, \|p\|^{\mathscr{M}_{\mathrm{mg}\mathscr{D}}})$$
$$\mathscr{M}_{\mathrm{mg}\mathscr{D}}, w \models p\Diamond\!\!\rightarrow_k q \text{ iff } \|q\|^{\mathscr{M}_{\mathrm{mg}\mathscr{D}}} \in \overline{g}_k(t, \|p\|^{\mathscr{M}_{\mathrm{mg}\mathscr{D}}}),$$

respectively. Its basic idea is also the same as in fuzzy-measure-based semantics for graded modal logic defined in [1, 2, 53].

Several soundness results based on belief- and plausibility-function-based semantics (cf. [1, 2, 53]) are shown in Table 5.4.

Here are two typical cases. First we define a probability distribution on $T$ by

$$\mathrm{pr}(t) \stackrel{\text{def}}{=} \begin{cases} \frac{1}{a}, & \text{if } t \in \|p_X\|^{\mathscr{M}_{\mathrm{mg}\mathscr{D}}}, \\ 0, & \text{otherwise,} \end{cases}$$

| No. | Transaction (movie) | AH | HM | pr |
|---|---|---|---|---|
| 1 | Secret people | 1 | | $\frac{1}{8}$ |
| 2 | Monte Carlo baby | 1 | | $\frac{1}{8}$ |
| 3 | Roman holiday | 1 | | $\frac{1}{8}$ |
| 4 | My fair lady | 1 | | $\frac{1}{8}$ |
| 5 | Breakfast at Tiffany's | 1 | 1 | $\frac{1}{8}$ |
| 6 | Charade | 1 | 1 | $\frac{1}{8}$ |
| 7 | Two for the road | 1 | 1 | $\frac{1}{8}$ |
| 8 | Wait until dark | 1 | 1 | $\frac{1}{8}$ |
| 9 | Days of wine and rose | | 1 | 0 |
| 10 | The great race | | 1 | 0 |
| 11 | The pink panther | | 1 | 0 |
| 12 | Sunflower | | 1 | 0 |
| 13 | Some like it hot | | | 0 |
| 14 | 12 Angry men | | | 0 |
| 15 | The apartment | | | 0 |
| | ...... | | | |
| 100 | Les aventuriers | | | 0 |

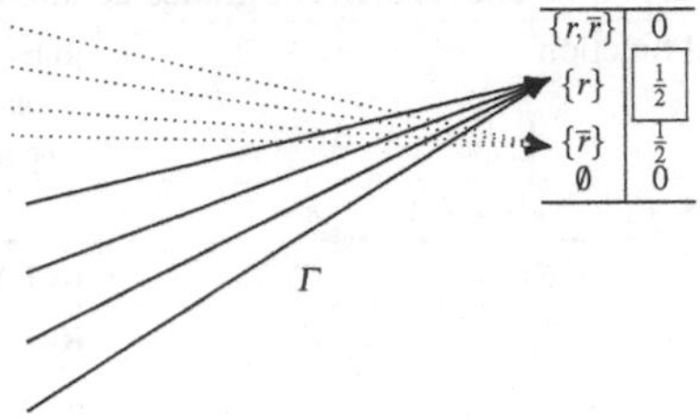

| | |
|---|---|
| $\{r, \overline{r}\}$ | 0 |
| $\{r\}$ | $\frac{1}{2}$ |
| $\{\overline{r}\}$ | $\frac{1}{2}$ |
| $\emptyset$ | 0 |

$\Gamma$

Fig. 5.1 An example of the strongest cases

where $a = |\|p_X\|^{\mathcal{M}_{\mathrm{mg}\mathcal{D}}}|$. This means that each world (transaction) $t$ in $\|p_X\|^{\mathcal{M}_{\mathrm{mg}\mathcal{D}}}$ is given an even mass (weight) $\frac{1}{a}$. To generalize the distribution is of course another interesting task.

Next, we shall see two typical cases of definition of $\Gamma$. First we describe strongest cases. When we define a mapping $\Gamma$ by

$$\Gamma(t) \stackrel{\mathrm{def}}{=} \begin{cases} \{r\}, & \text{if } t \in \|p_X\|^{\mathcal{M}_{\mathrm{mg}\mathcal{D}}}, \\ \{\overline{r}\}, & \text{otherwise.} \end{cases}$$

This means that the transactions in $\|p_X \wedge p_Y\|^{\mathcal{M}_{\mathrm{mg}\mathcal{D}}}$ contribute as evidence to $r$, while the transactions in $\|p_X \wedge \neg p_Y\|^{\mathcal{M}_{\mathrm{mg}\mathcal{D}}}$ contribute as evidence to $\overline{r}$. This is the strongest interpretation of co-occurrences.

Then, we can compute:

$$\mathrm{Bel}(r) = \frac{1}{a} \times b \text{ and } \mathrm{Pl}(r) = \frac{1}{a} \times b,$$

where $b = |\|p_X \wedge p_Y\|^{\mathcal{M}_{\mathrm{mg}\mathcal{D}}}|$. Thus the induced belief and plausibility functions collapse to the same probability measure pr: $\mathrm{Bel}(r) = \mathrm{Pl}(r) = \mathrm{pr}(r) = \frac{b}{a}$, and thus

$$c(r) = \left\langle \frac{b}{a}, \frac{b}{a} \right\rangle.$$

Hence this case represents the usual confidence. According to this idea, in our movie database, we can define pr and $\Gamma$ in the way in Fig. 5.1.

That is, any movie in $\|\mathrm{AH} \wedge \mathrm{HM}\|^{\mathcal{M}_{\mathrm{mg}\mathcal{D}}}$ contributes as evidence to that the rule holds ($r$), while all movie in $\|\mathrm{AH} \wedge \neg\mathrm{HM}\|^{\mathcal{M}_{\mathrm{mg}\mathcal{D}}}$ contributes as evidence to that the rule does not hold ($\overline{r}$). Thus we have

| No. | Transaction (movie) | AH | HM | pr |
|---|---|---|---|---|
| 1 | Secret people | 1 | | $\frac{1}{8}$ |
| 2 | Monte Carlo baby | 1 | | $\frac{1}{8}$ |
| 3 | Roman holiday | 1 | | $\frac{1}{8}$ |
| 4 | My fair lady | 1 | | $\frac{1}{8}$ |
| 5 | Breakfast at Tiffany's | 1 | 1 | $\frac{1}{8}$ |
| 6 | Charade | 1 | 1 | $\frac{1}{8}$ |
| 7 | Two for the road | 1 | 1 | $\frac{1}{8}$ |
| 8 | Wait until dark | 1 | 1 | $\frac{1}{8}$ |
| 9 | Days of wine and rose | | 1 | 0 |
| 10 | The great race | | 1 | 0 |
| 11 | The pink panther | | 1 | 0 |
| 12 | Sunflower | | 1 | 0 |
| 13 | Some like it hot | | | 0 |
| 14 | 12 Angry men | | | 0 |
| 15 | The apartment | | | 0 |
| | ...... | | | |
| 100 | Les aventuriers | | | 0 |

| | |
|---|---|
| $\{r,\overline{r}\}$ | $\frac{1}{2}$ |
| $\{r\}$ | 0 |
| $\{\overline{r}\}$ | $\frac{1}{2}$ |
| $\emptyset$ | 0 |

$\Gamma$

**Fig. 5.2** An example of the weakest cases

$$(\{AH\} \Longrightarrow \{HM\}) = \langle 0.5, 0.5\rangle.$$

Next, we describe weakest cases. In general, co-occurrences do not necessarily mean actual association. The weakest interpretation of co-occurrences is to consider transactions totally unknown as described as follows: When we define a mapping $\Gamma$ by $\Gamma(t) \stackrel{\text{def}}{=} \begin{cases} \{r, \overline{r}\}, \text{ if } t \in \|p_X\|^{\mathcal{M}_{\text{mg}\mathcal{D}}}, \\ \{\overline{r}\}, \quad \text{otherwise.} \end{cases}$
This means that the transactions in $\|p_X \wedge p_Y\|^{\mathcal{M}_{\text{mg}\mathcal{D}}}$ contribute as evidence to $R = \{r, \overline{r}\}$, while the transactions in $\|p_X \wedge \neg p_Y\|^{\mathcal{M}_{\text{mg}\mathcal{D}}}$ contribute as evidence to $\overline{r}$. Then, we can compute $\text{Bel}(r) = 0$ and $\text{Pl}(r) = \frac{1}{a} \times b$, and thus

$$c(r) = \left\langle 0, \frac{b}{a}\right\rangle.$$

According to this idea, in our movie database, we can define pr and $\Gamma$ in the way in Fig. 5.2.

That is, all movie in $\|\text{AH} \wedge \neg\text{HM}\|^{\mathcal{M}_{\text{mg}\mathcal{D}}}$ contributes as evidence to that the rule does not hold ($\overline{r}$), while we cannot expect whether each movie in $\|\text{AH} \wedge \text{HM}\|^{\mathcal{M}_{\text{mg}\mathcal{D}}}$ contributes or not as evidence to that the rule holds ($r$). Thus we have

$$c(\{AH\} \Longrightarrow \{HM\}) = \langle 0, 0.5\rangle.$$

In the case, the induced belief and plausibility functions, denoted respectively $\text{Bel}_{bpa'}$ and $\text{Pl}_{bpa'}$, become *necessity* and *possibility* measures in the sense of Dubois and Prade [54].

We have several soundness results based on necessity- and possibility-measure-based semantics (cf. [1, 2, 53]) shown in Table 5.5.

Finally, we describe an example of general cases. In the previous two typical cases, one of which coincides to the usual confidence, any transaction in $\|\text{AH} \wedge \text{HM}\|^{\mathcal{M}_{\text{mg}\mathcal{D}}}$

**Table 5.5** Soundness results of graded conditionals by necessity and possibility measures

| Necessity measure $0 < k \leq 1$ | Rules and axiom schemata | Possibility measure $0 < k \leq 1$ |
|---|---|---|
| ○ | **RCEA** | ○ |
| ○ | **RCEC** | ○ |
| ○ | **RCM** | ○ |
| ○ | **RCR** | |
| ○ | **RCN** | ○ |
| ○ | **RCK** | |
| ○ | **CM** | ○ |
| ○ | **CC** | |
| | **CF** | ○ |
| ○ | **CR** | ○ |
| ○ | **CN** | ○ |
| ○ | **CP** | ○ |
| ○ | **CK** | |
| ○ | **CD** | |
| | **CD**$_{\mathbf{C}}$ | ○ |

(or in $\|AH \wedge \neg HM\|^{\mathcal{M}_{mg\mathcal{D}}}$) has the same weight as evidence. It would be, however, possible that some of $\|AH{\wedge}HM\|^{\mathcal{M}_{mg\mathcal{D}}}$ (or $\|AH{\wedge}\neg HM\|^{\mathcal{M}_{mg\mathcal{D}}}$) does work as positive evidence to $r$ (or $\overline{r}$) but other part does not. Thus, we have a tool that allows us to introduce various kinds of 'a posteriori' pragmatic knowledge into the logical setting of association rules.

As an example, we assume that (1) the music of the first and second movies was not composed by Mancini, but the fact does not affect the validity of $\overline{r}$ because they are not very important ones, and (2) the music of the seventh movie was composed by Mancini, but the fact does not affect the validity of $r$.

Then we can define $\Gamma$ in the way in Fig. 5.3.

Thus, we have

$$c(\{AH\} \Longrightarrow \{HM\}) = \langle 0.375, 0.75 \rangle.$$

In general, users have such kind of knowledge 'a posteriori.' Thus the D-S based approach allows us to introduce various kinds of 'a posteriori' pragmatic knowledge into association rules.

In this section, we examined paraconsistency and paracompleteness that appear in association rules in a framework of probability-based models for conditional logics. For lower values of confidence (less than or equal to $\frac{1}{2}$), both $p \Box\!\!\rightarrow_k q$ and $p \Box\!\!\rightarrow_k \neg q$ may be generated so we must be careful to use such lower confidence.

Further, we extended the above discussion into the case of Dempster-Shafer theory of evidence to double-indexed confidences. Thus, the D-S based approach allows a sophisticated way of calculating confidence by introducing various kinds of 'a posteriori' pragmatic knowledge into association rules.

| No. | Transaction (movie) | AH | HM | pr |
|---|---|---|---|---|
| 1 | Secret people | 1 | | $\frac{1}{8}$ |
| 2 | Monte Carlo baby | 1 | | $\frac{1}{8}$ |
| 3 | Roman holiday | 1 | | $\frac{1}{8}$ |
| 4 | My fair lady | 1 | | $\frac{1}{8}$ |
| 5 | Breakfast at Tiffany's | 1 | 1 | $\frac{1}{8}$ |
| 6 | Charade | 1 | 1 | $\frac{1}{8}$ |
| 7 | Two for the road | 1 | 1 | $\frac{1}{8}$ |
| 8 | Wait until dark | 1 | 1 | $\frac{1}{8}$ |
| 9 | Days of wine and rose | | 1 | 0 |
| 10 | The great race | | 1 | 0 |
| 11 | The pink panther | | 1 | 0 |
| 12 | Sunflower | | 1 | 0 |
| 13 | Some like it hot | | | 0 |
| 14 | 12 Angry men | | | 0 |
| 15 | The apartment | | | 0 |
| | ...... | | | |
| 100 | Les aventuriers | | | 0 |

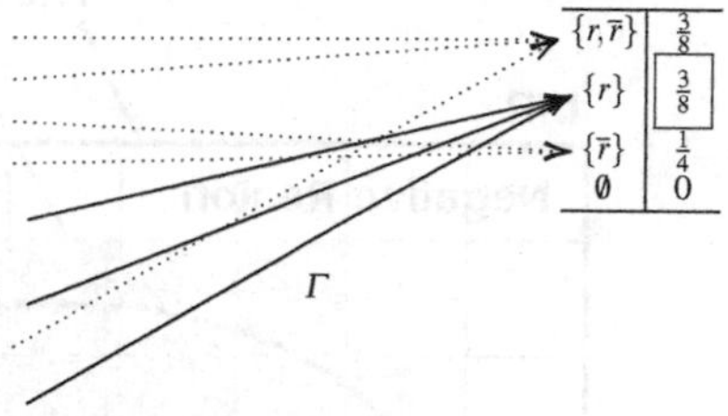

$\Gamma$

| | |
|---|---|
| $\{r, \bar{r}\}$ | $\frac{3}{8}$ |
| $\{r\}$ | $\frac{3}{8}$ |
| $\{\bar{r}\}$ | $\frac{1}{4}$ |
| $\emptyset$ | 0 |

**Fig. 5.3** An example of general cases

## 5.6 Background Knowledge in Reasoning

In this section, we examine some relationship between several kinds of reasoning processes and granularity generated by background knowledge based on Murai, Kudo and Akama [55]. We introduce two levels of objective and subjective under background knowledge.

In particular, we put much emphasis on the role of lower approximation, whose size depends on the granularity based on background knowledge, in several kinds of reasoning such as deduction, conflict resolution in expert systems and robot control.

Recently in Japan, *Kansei engineering* provides very interesting and important applications of rough set theory. There the concept of reducts plays an important part in such applications. We expect this section would give Kansei community another aspect of rough set theory, that is, adjustment of granularity.

Let $U$ be a universe of discourse and $R$ be an equivalence relation on $U$. In general, a relation on $U$ is a subset of the direct (Cartesian) product of $U$, i.e., $R \subseteq U \times U$. When a pair $(x, y)$ is in $R$, we write $xRy$.

A relation $R$ on $U$ is said to be an equivalence relation just in case it satisfies the following three properties: for every $x, y, z \in U$,

(1) $xRx$ (reflexivity)
(2) $xRy \Rightarrow yRx$ (symmetry)
(3) $xRy$ and $yRz \Rightarrow xRz$ (transitivity)

The set $[x]_R$ defined by

$$[x]_R = \{y \in U \mid xRy\}$$

is called the *equivalence class* of $x$ with respect to $R$.

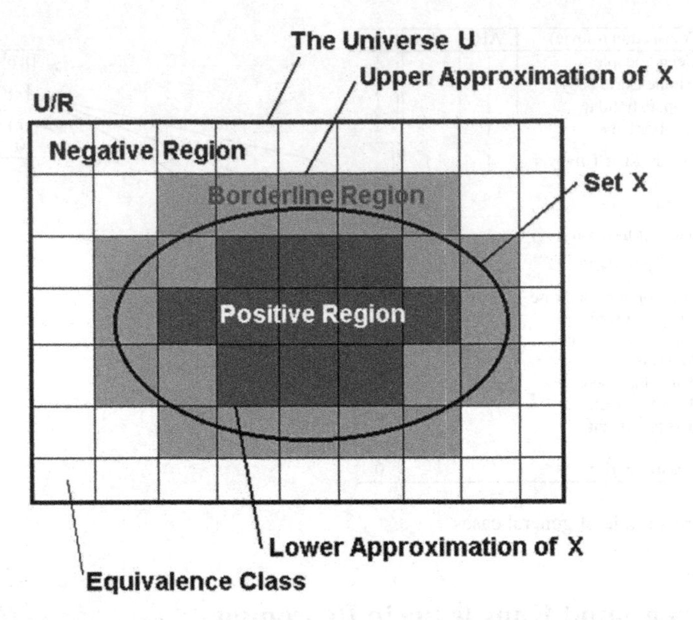


**Fig. 5.4** Two kinds of approximations

The family of all equivalence classes of each element in $U$ with respect to $R$ is denoted $U/R$, that is,

$U/R = \{[x]_R \mid x \in U\}$. It is called the *quotient set* of $U$ with respect to $R$.

Equivalence classes satisfy the following properties:

(1) $xRy \Rightarrow [x]_R = [y]_R$,
(2) $not(xRy) \Rightarrow [x]_R \cap [y]_R = \emptyset$.

Then, the quotient set $U/R$ gives a partition of $U$.

Thus, we can deal with equivalence classes as building blocks under background knowledge induced from relation $R$. In fact, we can approximate a set $X$ (unknown object) in the two ways just illustrated in Fig. 5.4.

One way is to make an approximation from inside using the building blocks of $U/R$ that are contained in $X$:

$\underline{R}(X) = \cup\{[x]_R \mid [x]_R \subseteq X\}$.

It is called the *lower approximation* of $X$ with respect to $R$. The other is to make an approximation from outside of $X$ by deleting the building blocks that have no intersection with $X$. Equivalently, it comes to

$\overline{R}(X) = \cup\{[x]_R \mid [x]_R \cap X \neq \emptyset\}$.

It is called the *upper approximation* of $X$ with respect to $R$. Obviously, we have the following inclusion:

$$\underline{R}(X) \subseteq X \subseteq \overline{R}(X).$$

Further, we use the following three terms:

(1) Positive region of $X$: $Pos(X) = \underline{R}(X)$
(2) Borderline region of $X$: $Bd(X) = \overline{R}(X) - \underline{R}(X)$
(3) Negative region of $X$: $Neg(X) = X - \overline{R}(X)$.

The pair

$$(\underline{R}(X), \overline{R}(X))$$

is called the *rough set* of $X$ with respect to $R$. And the pair

$$(U, R)$$

is referred as an *approximation* or *Pawlak space*.

Intuitively speaking, the size of building blocks depends on the granularity generated by a given approximation space or its quotient set. In Fig. 5.5, $U/R$ has coarser granularity that $U/R'$ has. Thus, in general, we can understand that $U/R'$ gives better approximation that $U/R$.

In order to deal with degrees of granularity in a quantitative way, several kinds of measures are introduced for the finite universe case. Among them, the following measure is called the *accuracy* of $X$ with respect to $R$:

$$\alpha_R(X) = |\underline{R}(X)| \,/\, |\overline{R}(X)|\,.$$

Another well-known measure is

$$\gamma_R(X) = |\underline{R}(X)| \,/\, |X|,$$

which is called the *quality* of $X$ with respect to $R$. By these measures we can have the degree of granularity of $X$ under background knowledge.

Now, we examine several kinds of reasoning under granularity generated from background knowledge. Compare the present discussion with the previous one.

Here, we note objective and subjective levels of knowledge. When fact $p$ is given, its proposition $P$ is just the maximum set of accessible worlds.

In ordinary reasoning, however, we cannot enumerate the total of them when carrying out reasoning processes. In general, we could imagine some proper subset of $P$ at most.

One possible way of specifying such subset is that we can consider some relevant worlds under background knowledge to be the lower approximation $\Box P$ of $P$. This is based on the idea that background knowledge formulates its own context with some granularity, in which our way of observing worlds is determined.

Size of lower approximation $\Box P$ depends on granularity generated by background knowledge. $P$ is in objective level while $\Box P$ in subjective level. There are several kinds of meaning of $\Box P$ in each context such as set of 'essential' or 'typical' elements.

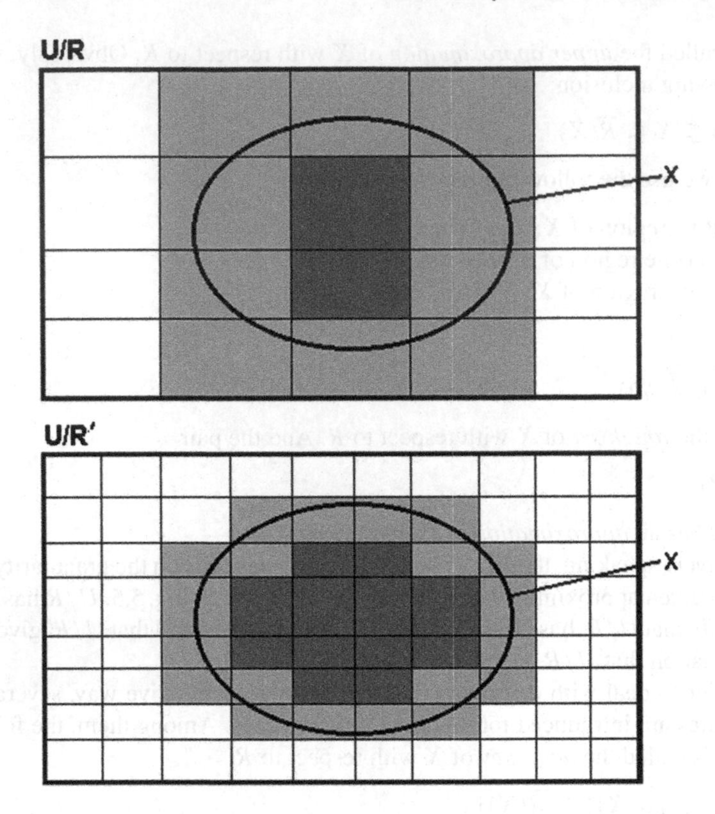


**Fig. 5.5** Adjusment of glanurarity

Logical reasoning in the usual sense, that is, *deduction* does not consider background knowledge. A typical rule of inference is well-known *modus ponens*:

$p, p \rightarrow q \; \Rightarrow \; q$ (From $p$ and $p \rightarrow q$ infer $q$).

This means that we can obtain a conclusion $q$ from a fact $p$ and rule $p \rightarrow q$. We examine modus ponens in the framework of possible world semantics.

Let $M = (U, R, v)$ be a Kripke model, where $U$ is a set of possible worlds, $R$ is an accessibility relation, and $v$ is a valuation function. Then, in $M$, rule $p \rightarrow q$ is represented as set inclusion between propositions:

$$M \models p \rightarrow q \; \Leftrightarrow \; P \subseteq Q.$$

Then the rule means the following procedure:

(1) Fact $p$ restricts the set of possible worlds that we can access under $p$.
(2) Then, by rule $p \rightarrow q$, we can find that conclusion $q$ is true at every world in the restricted set. That is, $q$ is necessary under fact $p$.

Thus, the rule implies the monotonicity of deduction. In fact we have

$$p \rightarrow q \; \Rightarrow \; \Box p \rightarrow \Box q$$

which holds in every Kripke models. We can rewrite it in a propositional level as follows (Fig. 5.6):
$P \subseteq Q \Rightarrow \Box P \subseteq \Box Q$.

Next, we discuss *non-monotonic reasoning*, which is one of most typical kind of ordinary reasoning. The Tweety example is well-known:

(1) Tweety is a bird.
Most birds fly.
Then he flies.

(2) Tweety is a penguin.
Penguins do not fly.
Then he does not fly.

Thus, the conclusion in (1) is withdrawn in (2).

Thus, the set of conclusions in non-monotonic reasoning no longer increase in a monotonic manner and in this sense the above kind of reasoning is said to be non-monotonic.

As stated before, the usual monotonic reasoning satisfies the monotonicity:

$$P \subseteq Q \Rightarrow \Box P \subseteq \Box Q.$$

while in non-monotonic reasoning in general,

$$P \not\subseteq Q, \text{ but } \Box P \subseteq \Box Q$$

In the Tweety case, let BIRD and FLYING be the set of birds and flying objects, respectively. Then in the objective level

$$\text{BIRD} \not\subseteq \text{FLYING}$$

but in the subjective level, the inclusion

$$\Box\text{BIRD} \subseteq \Box\text{FLYING}$$

holds (see Fig. 5.7).

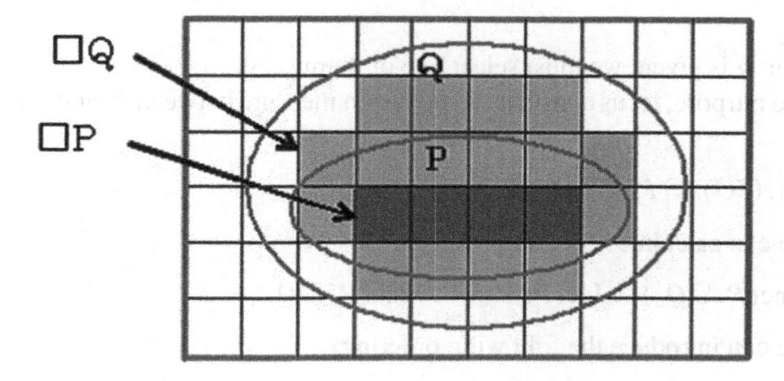


**Fig. 5.6** Deduction

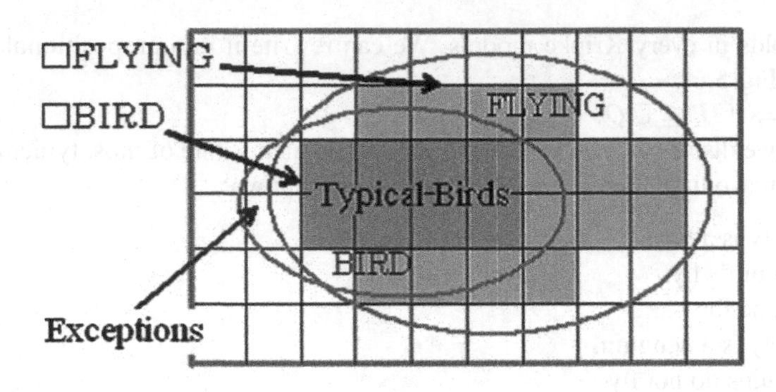


**Fig. 5.7** Non-Monotonic reasoning

We turn to *abduction*, which has the form of reasoning:

$$q, p \to q \Rightarrow p.$$

Apparently it is not valid because, in general, there are many sentences which imply $q$.

In 1883, Peirce introduced abduction as the inference of a case from a rule and a result; see Peirce [56]. He argued that abduction plays a very important role in scientific discovery in the 19th century.

Also abduction is important in human plausible reasoning, which is not necessary correct, like fortune-telling. When possible candidates of sentences which imply $q$ are given, we can give an order between the candidates using lower approximation of $Q$.

For example, in Fig. 5.8, we have three candidates, that is, we have three possible implications:

$$p_1 \to q,$$
$$p_2 \to q,$$
$$p_3 \to q.$$

Now when $q$ is given, we must select one of them.

For the purpose, let us consider the inclusion measure between $P_i$ and $\Box Q$ defined by

$$\mathrm{Inc}(P_i, \Box Q) = \mid P_i \cap \Box Q_i \mid / \mid P_i \mid.$$

Then, we can calculate

$$0 = \mathrm{Inc}(P, \Box Q_3) < \mathrm{Inc}(P, \Box Q_2) < \mathrm{Inc}(P, \Box Q_1) = 1$$

hence we can introduce the following ordering,

$$p_1 \geq p_2 \geq p_3.$$

Therefore, we can first choose $p_1$ as possible premise of abduction.

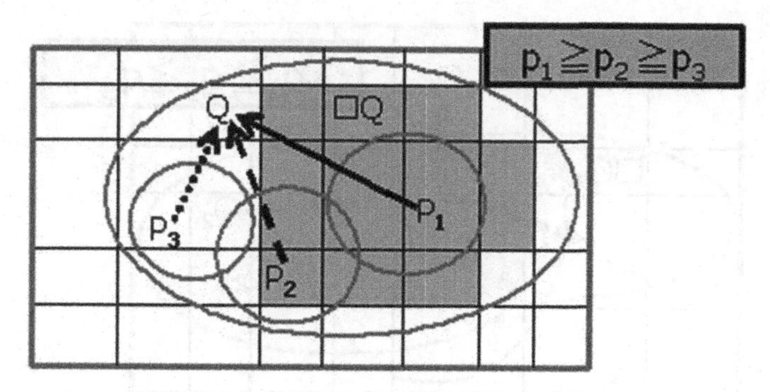


**Fig. 5.8** Abduction

Note that, in the example, the following implications with modality hold:

$$p_1 \to \Box q,$$
$$p_2 \to q,$$
$$p_3 \to (q \wedge \neg\Box q).$$

We can see the difference of strength of each possible premise for the same conclusion $q$.

The similar idea in abduction can be applied to *conflict resolution* in expert systems. Reasoning from an expert system may need conflict resolution to extract apropriate conclusions from conflicts.

We are given, for example, three monotonic rules:

$$p \to q_1,$$
$$p \to q_2,$$
$$p \to q_3.$$

Then, in a usual logical framework, we have conclusion

$$q_1 \wedge q_2 \wedge q_3$$

In many application areas, however, like expert systems and robot control, each conclusion is associated with action or execution, thus more than two conclusions cannot be carried out. Hence we must choose one of them (cf. Fig. 5.9).

Again let us consider the inclusion measure between $P$ and $\Box Q_i$ defined by

$$\mathrm{Inc}(P, \Box Q_i) = \mid P \cap \Box Q_i \mid / \mid P \mid .$$

Then, we can calculate

$$0 = \mathrm{Inc}(P, \Box Q_3) < \mathrm{Inc}(P, \Box Q_2) < \mathrm{Inc}(P, \Box Q_1) = 1$$

hence we can introduce the following ordering, $q_1 \geq q_2 \geq q_3$.
Therefore, it is plausible that we first select $q_1$ as possible conclusion with execution.

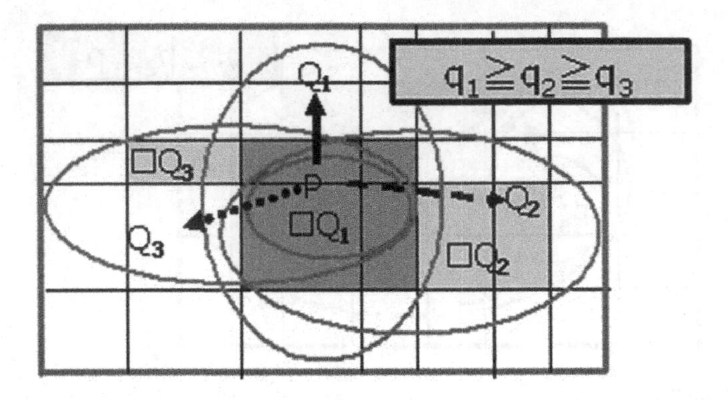


**Fig. 5.9** Conflict resolution

Note that, in the example, the following implications with modality hold:

$$p_1 \to \Box q_1,$$
$$p_2 \to q_2,$$
$$p_3 \to (q_3 \wedge \neg\Box q_3).$$

We can see the difference of strength of each possible conclusion for the same premise $p$.

We have shown an important role of background knowledge and granularity in several kinds of reasoning such as (monotonic) deduction, non-monotonic reasoning, abduction, and conflict resolution.

Thereby we have a foundation of applying Kansei representation into reasoning. Adjustment of granularity with topology could provide us another important characteristic of rough set theory in Kansei engineering as well as reducts.

Rough set theory has a close relationship with topological spaces [57] and adjustment of granularity can be regarded as homomorphism between topological spaces. Thus, we may introduce several useful concepts in topological spaces into Kansei engineering via rough set theory.

It is worth exploring rough set theory in the context of Kansei engineering. Also important is to implement a Kansei reasoning system and its application to recommendation systems [58] and image retrieval systems [59].

Kudo et al. [58] proposed a simple recommendation system based on rough set theory. Their recommendation method constructs decision rules from user's query and recommends some products by estimating implicit conditions of products based on decision rules.

Murai et al. [59] examined a logical representation of images by means of multi-rough sets. They introduced a level of granularization into a color space and then defined approximations of images. Further, they extended the idea to conditionals in images, which is expected to give a basis of image indexing and retrieval by images themselves.

# References

1. Murai, T., Miyakoshi, M., Shinmbo, M.: Measure-based semantics for modal logic. In: Lowen, R., Roubens, M. (eds.) Fuzzy Logic: State of the Arts, pp. 395–405. Kluwer, Dordrecht (1993)
2. Murai, T., Miyakoshi, M., Shimbo,M.: Soundness and completeness theorems between the Dempster-Shafer theory and logic of belief. In: Proceedings of the 3rd FUZZ-IEEE (WCCI), pp. 855–858 (1994)
3. Minsky, M.: A framework for representing knowledge. In: Haugeland, J. (ed.), Mind-Design, pp. 95–128. MIT Press, Cambridge, Mass, (1975)
4. McDermott, D., Doyle, J.: Non-monotonic logic I. Artif. Intell. **13**, 41–72 (1980)
5. McDermott, D.: Nonmonotonic logic II. J. ACM **29**, 33–57 (1982)
6. Marek, W., Shvartz, G., Truszczynski, M.: Modal nonmonotonic logics: ranges, characterization, computation. In: Proceedings of KR'91, pp. 395–404 (1991)
7. Reiter, R.: A logic for default reasoning. Artif. Intell. **13**, 81–132 (1980)
8. Etherington, D.W.: Reasoning with Incomplete Information. Pitman, London (1988)
9. Lukasiewicz, W.: Considerations on default logic. Comput. Intell. **4**, 1–16 (1988)
10. Lukasiewicz, W.: Non-Monotonic Reasoning: Foundation of Commonsense Reasoning. Ellis Horwood, New York (1992)
11. Besnard, P.: Introduction to Default Logic. Springer, Berlin (1989)
12. Moore, R.: Possible-world semantics for autoepistemic logic. In: Proceedings of AAAI Non-Monotonic Reasoning Workshop, pp. 344–354 (1984)
13. Moore, R.: Semantical considerations on nonmonotonic logic. Artif. Intell. **25**, 75–94 (1985)
14. Hintikka, S.: Knowledge and Belief. Cornell University Press, Ithaca (1962)
15. Stalnaker, R.: A note on non-monotonic modal logic. Artifi. Intell. **64**, 183–1963 (1993)
16. Konolige, K.: On the relation between default and autoepistemic logic. Artifi. Intell. **35**, 343–382 (1989)
17. Levesque, H.: All I know: a study in autoepistemic logic. Artifi. Intell. **42**, 263–309 (1990)
18. McCarthy, J.: Circumscription—a form of non-monotonic reasoning. Artifi. Intell. **13**, 27–39 (1980)
19. McCarthy, J.: Applications of circumscription to formalizing commonsense reasoning. Artifi. Intell. **28**, 89–116 (1984)
20. Lifschitz, V.: Computing circumscription. In: Proceedings of IJCAI'85, pp. 121–127 (1985)
21. Gabbay, D.M.: Theoretical foundations for non-monotonic reasoning in expert systems. In: Apt, K.R. (ed.) Logics and Models of Concurrent Systems, pp. 439–459. Springer (1984)
22. Makinson, D.: General theory of cumulative inference. In: Proceedings of the 2nd International Workshop on Non-Monotonic Reasoning, pp. 1–18. Springer (1989)
23. Makinson, D.: General patterns in nonmonotonic reasoning. In: Gabbay, D., Hogger, C., Robinson, J.A. (eds.) Handbook of Logic in Artificial Intelligence and Logic Programming, vol. 3, pp. 25–110. Oxford University Press, Oxford (1994)
24. Shoham, Y.: A semantical approach to nonmonotonic logics. Proc. Logic Comput. Sci., 275–279 (1987)
25. Kraus, S., Lehmann, D., Magidor, M.: Non-monotonic reasoning, preference models and cumulative reasoning. Artifi. Intell. **44**, 167–207 (1990)
26. Lehmann, D., Magidor, M.: What does a conditional knowledge base entail? Artifi. Intell. **55**(1992), 1–60 (1992)
27. Codd, E.: A relational model of data for large shared data banks. Commun. ACM **13**, 377–387 (1970)
28. Kowalski, R.: Predicate logic as a programming language. In: Proceedings of IFIP'74, pp. 569–574 (1974)
29. Kowalski, R.: Logic for Problem Solving. North-Holland, Amsterdam (1979)
30. Robinson, J.A.: A machine-oriented logic based on the resolution principle. J. ACM **12**, 23–41 (1965)

31. Colmerauer, A., Kanoui, H., Pasero, R., Roussel, P.: Un Systeme de Comunication Homme-machine en Fracais. Universite d'Aix Marseille (1973)
32. van Emden, M., Kowalski, R.: The semantics of predicate logic as a programming language. J. ACM **23**, 733–742 (1976)
33. Stoy, J.: Denotational Semantics: The Scott-Strachey Approach to Programming Language Theory. MIT Press, Cambridge Mass (1977)
34. Apt, K.R., van Emden, M.H.: Contributions to the theory of logic programming. J. ACM **29**, 841–862 (1982)
35. Clark, K.: Negation as failure. In: Gallaire, H., Minker, J. (eds.) Logic and Data Bases, pp. 293–322. Plenum Press, New York (1978)
36. Jaffar, J., Lassez, J.-L., Lloyd, J.: Completeness of the negation as failure rule. In: Proceedings of IJCAI'83, pp. 500–506 (1983)
37. Lloyd, J.: Foundations of Logic Programming. Springer, Berlin (1984)
38. Lloyd, J.: Foundations of Logic Programming, 2nd edn. Springer, Berlin (1987)
39. Reiter, R.: On closed world data bases. In: Gallaire, H., Minker, J. (eds.) Logic Data Bases, pp. 55–76. Plenum Press, New York (1978)
40. Minker, J.: On indefinite deductive databases and the closed world assumption. In: Loveland, D. (ed.), Proceedings of the 6th International Conference on Automated Deduction, pp. 292–308. Springer, Berlin (1982)
41. Apt, K., Blair, H., Walker, A.: Towards a theory of declarative knowledge. In: Minker, J. (ed.), Foundations of Deductive Databases and Logic Programming, pp. 89–148. Morgan Kaufmann, Los Altos (1988)
42. Gelfond, M., Lifschitz, V.: The stable model semantics for logic programming. In: Proceedings of ICLP'88, pp. 1070–1080 (1988)
43. Van Gelder, A., Ross, K., Schipf, J.: The well-founded semantics for general logic programs. J. ACM **38**, 620–650 (1991)
44. Gelfond, M., Lifschitz, V.: Logic programs with classical negation. In: Proceedings of ICLP'90, pp. 579–597 (1990)
45. Kowalski, R., Sadri, F.: Logic programs with exceptions. In: Proceeding of ICLP'90, pp. 598–613 (1990)
46. Abe, J.M., Akama, S., Nakamatsu, K.: Introduction to Annotated Logics. Springer, Heidelberg (2015)
47. Chellas, B.: Modal Logic: An Introduction. Cambridge University Press, Cambridge (1980)
48. Agrawal, R., Imielinski, T., Swami, A.: Mining association rules between sets of items in large databases. In: Proceedings of the ACM SIGMOD Conference on Management of Data, pp. 207–216 (1993)
49. Dempster, A.P.: Upper and lower probabilities induced by a multivalued mapping. Ann. Math. Stat. **38**, 325–339 (1967)
50. Shafer, G.: A Mathematical Theory of Evidence. Princeton University Press, Princeton (1976)
51. Lewis, D.: Counterfactuals. Blackwell, Oxford (1973)
52. Murai, T., Sato, Y., Kudo,Y.: Paraconsistency and neighborhood models in modal logic. In: Proceedings of the 7th World Multiconference on Systemics, Cybernetics and Informatics, vol. XII, pp. 220–223 (2003)
53. Murai, T., Miyakoshi, M., Shinmbo, M.: A logical foundation of graded modal operators defined by fuzzy measures. In: Proceedings of the 4th FUZZ-IEEE, pp. 151–156, Kluwer, Dordrecht (1995); Semantics for modal logic. Fuzzy Logic State of the Arts, 395–405 (1993)
54. Dubois, D., Prade, H.: Possibility Theory: An Approach to Computerized Processing of Uncertainty. Springer, Berlin (1988)
55. Murai, T., Kudo, Y., Akama, S.: Towards a foundation of Kansei representation in human reasoning. Kansei Eng. Int. **6**, 41–46 (2006)
56. Peirce, C: Collected Papers of Charles Sanders Peirce, vol. 8. In: Hartshone, C., Weiss, P., Burks, A. (eds.) Harvard University Press, Cambridge, MA (1931–1936)
57. Munkers, J.: Topology, 2nd edn. Prentice Hall, Upper Saddle River, NJ (2000)

58. Kudo, Y., Amano, S., Seino, T., Murai, T.: A simple recommendation system based on rough set theory. Kansei Eng. **6**, 19–24 (2006)
59. Murai, T., Miyamoto, S., Kudo, Y.: A logical representation of images by means of multi-rough sets for Kansei image retrieval. In: Proceedings of RSKT 2007, pp. 244–251. Springer, Heidelberg (2007)

# Chapter 6
# Conclusions

**Abstract** This chapter gives some conclusions with the summary of the book. We evaluate our work in connection with others. We also discuss several issues to be investigated.

## 6.1 Summary

Rough set theory is very suitable to model imprecise and uncertain data, and it has many applications. Clearly, it is also applied to model general reasoning. In this book, we developed a granularity-based framework for reasoning based on modal logic. This can characterize deduction, induction and abduction within the possible world semantics for modal logic.

There are many logic-based approaches to reasoning in the literature. We believe that rough set theory can serve as a promising framework. Our approach has a firm foundation, namely a generalization of the approach to rough set theory based on modal logic.

We have introduced an $\alpha$-level fuzzy measure model based on background knowledge and proposed a unified formulation of deduction, induction, and abduction based on this model. Using the proposed model, we have characterized typical situations of the given facts and rules by $(1 - \alpha)$-lower approximation of truth sets of non-modal sentences that represent the given facts and rules.

We have also proven that the modal system EMND45 is sound with respect to the class of all $\alpha$-level fuzzy measure models based on background knowledge. Moreover, we have characterized deduction, induction, and abduction as reasoning processes based on typical situations.

In the proposed framework, deduction and abduction are illustrated as valid reasoning processes based on typical situations of facts. On the other hand, induction is illustrated as a reasoning process of generalization based on observations. This means that our framework can properly model resoning processes for deduction, induction and abduction.

Furthermore, in the $\alpha$-level fuzzy measure model based on background knowledge, we have pointed out that induction has non-monotonicity based on revision of



S. Akama et al., *Reasoning with Rough Sets*, Intelligent Systems Reference Library 142, https://doi.org/10.1007/978-3-319-72691-5_6

the indiscernibility relation in the given Kripke model as background knowledge and gave an example in which a rule inferred by induction based on typical situations is rejected by refinement of the indiscernibility relation.

Our approach can provide a novel and generic framework for reasoning based on modal logic, which is derived from previous modal logic approaches. This means that standard non-classical logic, i,e., modal logic, is attractive without developing a new theory of non-monotonicity.

We can characterize deduction, induction and abduction in a single framework. It is difficult to deal with these three types of reasoning in a unified manner. For example, classical logic is appropriate to deduction, but it fails to model induction and abduction. Since classical logic was developed to formalize mathematical reasoning, it is not surprising.

There are other types of logics for induction and abduction. A fuzzy (or probabilistic) formulation is attractive for induction. For abduction, we generally have to advance a specific formalization for abduction, i.e., abductive logic. Logics for induction and abduction can be developed from different persprctives, and there is in fact a rich literature.

We can also accommodate to non-monotonicity which is fundamental to commonsense reasoning. It is known that the so-called non-monotonic logic was developed to formalize non-monotonic reasoning. Our approach revealed that we can dispense with non-standard logics like non-monotonic logics.

We have discussed the role of background knowledge for various types of reasoning. We have introduced two levels of objective and subjective under background knowledge. The use of these levels depends on the applications under consideration. We have also addressed the role of background knowledge for Kansei engineering.

We believe that rough set theory is of special use for providing a unified basis for reasoning. In fact, our granularity-based framework offers one of such formulations, which has several advantages mentioned above.

## 6.2 Further Issues

The work presented in this book established the foundations for general reasoning based on rough set theory. It can properly handle various types of reasoning in AI and related areas and opens a number of theoretical and practical problems which should be addressed in the future research.

One of the most important directions is the treatment of iteration of deduction, induction, and abduction in the proposed framework. All reasoning processes we have treated in this book are one-step reasoning in that the reasoning processes are complete after using either deduction, induction, or abduction only once. However, for real applications, we must deal with iteration of deduction, induction, and abduction.

Therefore, we need to extend the proposed framework to treat multiple-step reasoning, and we think that this extension will be closely connected to *belief revision* [1, 2] and *belief update* [3] in the proposed framework.

Segerberg [4] proposed *dynamic doxastic logic* for belief revision. His logic can describe belief revision in the object level as opposed to the orginal formulation of belief revision. Thus, it would be further possible to incorporate fuzzy (or probabilistic) measures into Segerberg's logic.

Notice that choice of the degree $\alpha \in (0.5, 1]$ affects the results of reasoning directly in the sense of whether and to what degree we allow the existence of exceptions to typical situations. However, we assumed that $\alpha$ is given in this book, and we have not discussed how to choose $\alpha$. Considering and introducing some criteria for choosing $\alpha$ are also important for the proposed framework.

We have used modal logics based on $\alpha$-level fuzzy measure model for our framework, and showed their axiomatizations. Although we did not show their completeness, it is interesting to prove it. For applications, we need to work out practical proof methods using sequent and tableau calculi.

Our modal logics can cope with fuzziness, and have intimate connections with related logics, e.g., graded modal logics. Therefore, we need to clarify precise relationships between our modal logic and related ones.

The presented framework can accommodate to paraconsistency and paracompleteness. This implies that it can also formalize incomplete and inconsistent information in intelligent systems. In formal logic, several non-classical logics for such concepts have been developed. Unfortunately, relations to such logics are not obvious, and it is necessary to clarify the relations.

Although our framework is based on possible worlds semantics for modal logic, it is interesting to explore alternative foundations. There are some candidates for this purpose. Orlowska's logic for reasoning about knowledge is very powerful for knowledge representation, and we should further study it. In particular, we have to find its complete axiomatization with a proof theory. Various logics for knowledge representations should be also studied in relation to rough set theory.

For other non-classical logics, many-valued logic seems attractive. We know that the work in this direction has been done within three-valued logic as in Avron and Knonikowska's approach. We believe that their approach can be extended in various ways. The approach can be in fact expanded for four-valued logic and other many-valued logics.

We also note that it is also promising to investigate an alternative formulation of rough set logics on the basis of relevance logic and paraconsistent logic. Since relevance logic is inspired by the correct interpretation of implication, it is of help to the formulation of association rule. Paraconsistent logic may be useful to the treatments of inconsistency in rough set theory.

Recently, Lin proposed *granular computing* based on rough set theory to motivate a new type of computing in Lin [5]. This is obviously an interesting idea. It would be thus possible to apply the idea of granular computing into various (human) reasoning processes. The work in this line may be found in Murai et al. [6, 7]. It is clear that the approach is also closely related to Kansei engineering, and more work should be done.

We hope to tackle these issues in our future papers and books. We should also work out several applications using our framework in a number of areas from AI to engineering.

## References


1. Gärdenfors, P.: Knowledge in Flux: Modeling the Dynamics of Epistemic States. MIT Press, Cambridge, Mass (1988)
2. Katsuno, H., Mendelzon, A.: Propositional knowledge base revision and minimal change. Artif. Intell. **52**, 263–294 (1991)
3. Katsuno, H., Mendelzon, A.: On the difference between updating a knowledge base and revising it. In: Gärdenfors, P. (ed.) Belief Revision. Cambridge University Press, Cambridge (1992)
4. Segerberg, K.: Irrevocable belief revision in dynamic doxastic logic. Notre Dame J. Form. Logic **39**, 287–306 (1998)
5. Lin, T.: Granular computing on baniary relation, I and II. In: Polkowski et al. (eds.) Rough Sets in Knowledge Discovery pp. 107–121, 122–140. Physica-Verlag (1998)
6. Murai, T., Nakata, M., Sato: A note on filtration and granular resoning. In: Terano et al. (eds.) New Frontiers in Artificial Intelligence, LNAI 2253, pp. 385–389. (2001)
7. Murai, T., Resconi, G., Nakata, M. and Sato, Y.: Operations of zooming in and out on possible worlds for semantic fields, E. Damiani et al. (eds.), *Knowledge-Based Intelligent Information Engineering Systems and Allied Technology*, 1083–1087, IOS Press, 2002

# References


1. Abe, J.M.: On the foundations of annotated logics (in Portuguese). Ph.D. thesis, University of São Paulo, Brazil (1992)
2. Abe, J.M., Akama, S., Nakamatsu, K.: Introduction to Annotated Logics. Springer, Heidelberg (2015)
3. Adsiaans, P., Zantinge, D.: Data Mining. Addison-Wesley, Reading, Mass (1996)
4. Agrawal, R., Imielinski, T., Swami, A.: Mining association rules between sets of items in large databases. In: Proceedings of the ACM SIGMOD Conference on Management of Data, pp. 207–216 (1993)
5. Akama, S.: Resolution in constructivism. Log. et Anal. **120**, 385–399 (1987)
6. Akama, S.: Constructive predicate logic with strong negation and model theory. Notre Dame J. Form. Log. **29**, 18–27 (1988)
7. Akama, S.: On the proof method for constructive falsity. Z. für Math. Log. und Grundl. der Math. **34**, 385–392 (1988)
8. Akama, S.: Subformula semantics for strong negation systems. J. Philos. Log. **19**, 217–226 (1990)
9. Akama, S.: Constructive falsity: foundations and their applications to computer science. Ph.D. thesis, Keio University, Yokohama, Japan (1990)
10. Akama, S.: The Gentzen-Kripke construction of the intermediate logic LQ. Notre Dame J. Form. Log. **33**, 148–153 (1992)
11. Akama, S.: Nelson's paraconsistent logics. Log. Log. Philos. **7**, 101–115 (1999)
12. Akama, S., Murai, T.: Rough set semantics for three-valued logics. In: Nakamatsu, K., Abe, J.M. (eds.) Advances in Logic Based Intelligent Systems, pp. 242–247. IOS Press, Amsterdam (2005)
13. Akama, S., Murai, T. and Kudo, Y.: Heyting-Brouwer rough set logic. In: Proceedings of KSE2013, Hanoi, pp. 135–145. Springer, Heidelberg (2013)
14. Akama, S., Murai, T., Kudo, Y.: Da Costa logics and vagueness. In: Proceedings of GrC2014, Noboribetsu, Japan (2014)
15. Almukdad, A., Nelson, D.: Constructible falsity and inexact predicates. J. Symb. Log. **49**, 231–233 (1984)



S. Akama et al., *Reasoning with Rough Sets*, Intelligent Systems Reference Library 142, https://doi.org/10.1007/978-3-319-72691-5

16. Anderson, A., Belnap, N.: Entailment: The Logic of Relevance and Necessity I. Princeton University Press, Princeton (1976)
17. Anderson, A., Belnap, N., Dunn, J.: Entailment: The Logic of Relevance and Necessity II. Princeton University Press, Princeton (1992)
18. Apt, K., Blair, H., Walker, A.: Towards a theory of declarative knowledge. In: Minker, J. (ed.), Foundations of Deductive Databases and Logic Programming, pp. 89–148. Morgan Kaufmann, Los Altos (1988)
19. Arieli, O., Avron, A.: Reasoning with logical bilattices. J. Log. Lang. Inf. **5**, 25–63 (1996)
20. Arieli, O., Avron, A.: The value of fur values. Artif. Intell. **102**, 97–141 (1998)
21. Apt, K.R., van Emden, M.H.: Contributions to the theory of logic programming. J. ACM **29**, 841–862 (1982)
22. Armstrong, W.: Dependency structures in data base relationships. In: IFIP'74, pp. 580–583 (1974)
23. Arruda, A.I.: A survey of paraconsistent logic. In: Arruda, A., da Costa, N., Chuaqui, R. (eds.) Mathematical Logic in Latin America, pp. 1–41. North-Holland, Amsterdam (1980)
24. Atnassov, K.: Intuitionistic Fuzzy Sets. Physica, Haidelberg (1999)
25. Asenjo, F.G.: A calculus of antinomies. Notre Dame J. Form. Log. **7**, 103–105 (1966)
26. Avron, A., Konikowska, B.: Rough sets and 3-valued logics. Stud. log **90**, 69–92 (2008)
27. Avron, A., Lev, I.: Non-deterministic multiple-valued structures. J. Logic Comput. **15**, 241–261 (2005)
28. Balbiani, P.: A modal logic for data analysis. In: Proceedings of MFCS'96, LNCS 1113, pp. 167–179. Springer, Berlin.
29. Batens, D.: Dynamic dialectical logics. In: Priest, G., Routley, R., Norman, J. (eds.) Paraconsistent Logic: Essay on the Inconsistent, pp. 187–217. Philosophia Verlag, München (1989)
30. Batens, D.: Inconsistency-adaptive logics and the foundation of non-monotonic logics. Log. et Anal. **145**, 57–94 (1994)
31. Batens, D.: A general characterization of adaptive logics. Log. et Anal. **173–175**, 45–68 (2001)
32. Belnap, N.D.: A useful four-valued logic. In: Dunn, J.M., Epstein, G. (eds.) Modern Uses of Multi-Valued Logic, pp. 8–37. Reidel, Dordrecht (1977)
33. Belnap, N.D.: How a computer should think. In: Ryle, G. (ed.) Contemporary Aspects of Philosophy, pp. 30–55. Oriel Press (1977)
34. Besnard, P.: Introduction to Default Logic. Springer, Berlin (1989)
35. Bit, M., Beaubouef, T.: Rough set uncertainty for robotic systems. J. Comput. Syst. Coll. **23**, 126–132 (2008)
36. Blair, H.A., Subrahmanian, V.S.: Paraconsistent logic programming. Theor. Comput. Sci. **68**, 135–154 (1989)
37. Carnielli, W., Coniglio, M., Marcos, J.: Logics of formal inconsistency. In: Gabbay, D., Guenthner, F. (eds.) Handbook of Philosophical Logic, vol. 14, 2nd edn, pp. 1–93. Springer, Heidelberg (2007)

38. Carnielli, W., Marcos, J.: Tableau systems for logics of formal inconsistency. In: Abrabnia, H.R. (ed.) Proceedings of the 2001 International Conference on Artificial Intelligence, vol. II, pp. 848–852. CSREA Press (2001)
39. Chellas, B.: Modal Logic: An Introduction. Cambridge University Press, Cambridge (1980)
40. Clark, K.: Negation as failure. In: Gallaire, H., Minker, J. (eds.) Logic and Data Bases, pp. 293–322. Plenum Press, New York (1978)
41. Codd, E.: A relational model of data for large shared data banks. Commun. ACM **13**, 377–387 (1970)
42. Colmerauer, A., Kanoui, H., Pasero, R., Roussel, P.: Un Systeme de Comunication Homme-machine en Fracais. Universite d'Aix Marseille (1973)
43. Cornelis, C., De Cock, J., Kerre, E.: Intuitionistic fuzzy rough sets: at the crossroads of imperfect knowledge. Expert Syst. **20**, 260–270 (2003)
44. de Caro, F.: Graded modalities II. Stud. log **47**, 1–10 (1988)
45. da Costa, N.C.A.: $\alpha$-models and the system $T$ and $T^*$. Notre Dame J. Form. Log. **14**, 443–454 (1974)
46. da Costa, N.C.A.: On the theory of inconsistent formal systems. Notre Dame J. Form. Log. **15**, 497–510 (1974)
47. da Costa, N.C.A., Abe, J.M., Subrahmanian, V.S.: Remarks on annotated logic. Z. für Math. Log. und Grundl. der Math. **37**, 561–570 (1991)
48. da Costa, N.C.A., Alves, E.H.: A semantical analysis of the calculi $C_n$. Notre Dame J. Form. Log. **18**, 621–630 (1977)
49. da Costa, N.C.A., Subrahmanian, V.S., Vago, C.: The paraconsistent logic $\mathcal{PT}$. Z. für Math. Log. und Grundl. der Math. **37**, 139–148 (1991)
50. Dempster, A.P.: Upper and lower probabilities induced by a multivalued mapping. Ann. Math. Stat. **38**, 325–339 (1967)
51. Dubois, D., Prade, H.: Possibility Theory: An Approach to Computerized Processing of Uncertainty. Springer, Berlin (1988)
52. Dubois, D., Prade, H.: Rough fuzzy sets and fuzzy rough sets. Int. J. Gen. Syst. **17**, 191–209 (1989)
53. Dummett, M.: A propositional calculus with denumerable matrix. J. Symb. Log. **24**, 97–106 (1959)
54. Dunn, J.M.: Relevance logic and entailment. In: Gabbay, D., Gunthner, F. (eds.) Handbook of Philosophical Logic, vol. III, pp. 117–224. Reidel, Dordrecht (1986)
55. Düntsch, I.: A logic for rough sets. Theor. Comput. Sci. **179**, 427–436 (1997)
56. Etherington, D.W.: Reasoning with Incomplete Information. Pitman, London (1988)
57. Fagin, R., Halpern, J., Moses, Y., Vardi, M.: Reasoning About Knowledge. MIT Press, Cambridge, Mass (1995)
58. Fariñas del Cerro, L., Orlowska, E.: DAL-a logic for data analysis. Theor. Comput. Sci. **36**, 251–264 (1985)
59. Fattorosi-Barnaba, M., Amati, G.: Modal operators with probabilistic interpretations I. Stud. Log. **46**, 383–393 (1987)

60. Fattorosi-Barnaba, M., de Caro, F.: Graded modalities I. Stud. Log. **44**, 197–221 (1985)
61. Fattorosi-Barnaba, M., de Caro, F.: Graded modalities III. Stud. Log. **47**, 99–110 (1988)
62. Fitting, M.: Intuitionisic Logic, Model Theory and Forcing. North-Holland, Amsterdam (1969)
63. Fitting, M.: Bilattices and the semantics of logic programming. J. Log. Program. **11**, 91–116 (1991)
64. Fitting, M.: A theory of truth that prefers falsehood. J. Philos. Log. **26**, 477–500 (1997)
65. Gabbay, D.M.: Theoretical foundations for non-monotonic reasoning in expert systems. In: Apt, K.R. (ed.) Logics and Models of Concurrent Systems, pp. 439–459. Springer (1984)
66. Ganter, B., Wille, R.: Formal Concept Analysis. Springer, Berlin (1999)
67. Gärdenfors, P.: Knowledge in Flux: Modeling the Dynamics of Epistemic States. MIT Press, Cambridge, Mass (1988)
68. Gentzen, G.: Collected papers of Gerhard Gentzen. In: Szabo, M.E. (ed.). North-Holland, Amsterdam (1969)
69. Gelfond, M., Lifschitz, V.: The stable model semantics for logic programming. In: Proceedings of ICLP'88, pp. 1070–1080 (1988)
70. Gelfond, M., Lifschitz, V.: Logic programs with classical negation. In: Proceedings of ICLP'90, pp. 579–597 (1990)
71. Ginsberg, M.: Multivalued logics. In: Proceedings of AAAI 1986, pp. 243–247. Morgan Kaufman, Los Altos (1986)
72. Ginsberg, M.: Multivalued logics: a uniform approach to reasoning in AI. Comput. Intell. **4**, 256–316 (1988)
73. Halpern, J., Moses, Y.: Towards a theory of knowledge and ignorance: preliminary report. In: Apt, K. (ed.) Logics and Models of Concurrent Systems, pp. 459–476. Springer, Berlin (1985)
74. Halpern, J., Moses, Y.: A theory of knowledge and ignorance for many agents. J. Logic Comput. **7**, 79–108 (1997)
75. Heyting, A.: Intuitionism. North-Holland, Amsterdam (1952)
76. Hintikka, S.: Knowledge and Belief. Cornell University Press, Ithaca (1962)
77. Hirano, S., Tsumoto, S.: Rough representation of a region of interest in medical images. Int. J. Approx. Reason. **40**, 23–34 (2005)
78. Hughes, G. and Cresswell, M.: An Introduction to Modal Logic. Methuen, London (1968)
79. Hughes, G., Cresswell, M.: A New Introduction to Modal Logic. Routledge, New York (1996)
80. Iturrioz, L.: Rough sets and three-valued structures. In: Orlowska, E. (ed.) Logic at Work: Essays Dedicated to the Memory of Helena Rasiowa, pp. 596–603. Physica-Verlag, Heidelberg (1999)
81. Iwinski, T.: Algebraic approach to rough sets. Bull. Pol. Acad. Math. **37**, 673–683 (1987)

82. Jaffar, J., Lassez, J.-L., Lloyd, J.: Completeness of the negation as failure rule. In: Proceedings of IJCAI'83, pp. 500–506 (1983)
83. Järvinen, J., Pagliani, P., Radeleczki, S.: Information completeness in Nelson algebras of rough sets induced by quasiorders. Stud. Log. **101**, 1073–1092 (2013)
84. Jaśkowski, S.: Propositional calculus for contradictory deductive systems (in Polish). Stud. Soc. Sci. Tor. Sect. A **1**, 55–77 (1948)
85. Jaśkowski, S.: On the discursive conjunction in the propositional calculus for inconsistent deductive systems (in Polish). Stud. Soc. Sci. Tor. Sect. A **8**, 171–172 (1949)
86. Katsuno, H., Mendelzon, A.: Propositional knowledge base revision and minimal change. Artif. Intell. **52**, 263–294 (1991)
87. Katsuno, H., Mendelzon, A.: On the difference between updating a knowledge base and revising it. In: Gärdenfors, P. (ed.) Belief Revision. Cambridge University Press, Cambridge (1992)
88. Kifer, M., Subrahmanian, V.S.: On the expressive power of annotated logic programs. In: Proceedings of the 1989 North American Conference on Logic Programming, pp. 1069–1089 (1989)
89. Kleene, S.: Introduction to Metamathematics. North-Holland, Amsterdam (1952)
90. Konikowska, B.: A logic for reasoning about relative similarity. Stud. Log. **58**, 185–228 (1997)
91. Konolige, K.: On the relation between default and autoepistemic logic. Artif. Intell. **35**, 343–382 (1989)
92. Kotas, J.: The axiomatization of S. Jaskowski's discursive logic. Stud. Log. **33**, 195–200 (1974)
93. Kowalski, R.: Predicate logic as a programming language. In: Proceedings of IFIP'74, pp. 569–574 (1974)
94. Kowalski, R.: Logic for Problem Solving. North-Holland, Amsterdam (1979)
95. Kowalski, R., Sadri, F.: Logic programs with exceptions. In: Proceeding of ICLP'90, pp. 598–613 (1990)
96. Kraus, S., Lehmann, D., Magidor, M.: Non-monotonic reasoning, preference models and cumulative reasoning. Artif. Intell. **44**, 167–207 (1990)
97. Krisel, G., Putnam, H.: Eine unableitbarkeitsbeuwesmethode für den intuitinistischen Aussagenkalkul. Arch. für Math. Logik und Grundlagenforschung **3**, 74–78 (1967)
98. Kripke, S.: A complete theorem in modal logic. J. Symb. Log. **24**, 1–24 (1959)
99. Kripke, S.: Semantical considerations on modal logic. Acta Philos. Fenn. **16**, 83–94 (1963)
100. Kripke, S.: Semantical analysis of modal logic I. Z. für math. Logik und Grundl. der Math. **8**, 67–96 (1963)
101. Kripke, S.: Semantical analysis of intuitionistic logic. In: Crossley, J., Dummett, M. (eds.) Formal Systems and Recursive Functions, pp. 92–130. North-Holland, Amsterdam (1965)
102. Kripke, S.: Outline of a theory of truth. J. Philos. **72**, 690–716 (1975)

103. Kudo, Y., Murai, T., Akama, S.: A granularity-based framework of deduction, induction, and abduction. Int. J. Approx. Reason. **50**, 1215–1226 (2009)
104. Kudo, Y., Amano, S., Seino, T., Murai, T.: A simple recommendation system based on rough set theory. Kansei Eng. **6**, 19–24 (2006)
105. Lewis, D.: Counterfactuals. Blackwell, Oxford (1973)
106. Lehmann, D., Magidor, M.: What does a conditional knowledge base entail? Artif. Intell. **55**, 1–60 (1992)
107. Levesque, H.: All I know: a study in autoepistemic logic. Artif. Intell. **42**, 263–309 (1990)
108. Liau, C.-J.: An overview of rough set semantics for modal and quantifier logics. Int. J. Uncertain. Fuzziness Knowl. -Based Syst. **8**, 93–118 (2000)
109. Lifschitz, V.: Computing circumscription. In: Proceedings of IJCAI'85, pp. 121–127 (1985)
110. Lin, T.: Granular computing on baniary relation, I and II. In: Polkowski et al. (eds.) Rough Sets in Knowledge Discovery pp. 107–121, 122–140. Physica-Verlag (1998)
111. Lin, T., Cercone, N. (eds.): Rough Sets and Data Mining. Springer, Berlin (1997)
112. Lloyd, J.: Foundations of Logic Programming. Springer, Berlin (1984)
113. Lloyd, J.: Foundations of Logic Programming, 2nd edn. Springer, Berlin (1987)
114. Łukasiewicz, J.: On 3-valued logic 1920. In: McCall, S. (ed.) Polish Logic, pp. 16–18. Oxford University Press, Oxford (1967)
115. Łukasiewicz, J.: Many-valued systems of propositional logic, 1930. In: McCall, S. (ed.) Polish Logic. Oxford University Press, Oxford (1967)
116. Lukasiewicz, W.: Considerations on default logic. Comput. Intell. **4**, 1–16 (1988)
117. Lukasiewicz, W.: Non-Monotonic Reasoning: Foundation of Commonsense Reasoning. Ellis Horwood, New York (1992)
118. Makinson, D.: General theory of cumulative inference. In: Proceedings of the 2nd International Workshop on Non-Monotonic Reasoning, pp. 1–18. Springer (1989)
119. Makinson, D.: General patterns in nonmonotonic reasoning. In: Gabbay, D., Hogger, C., Robinson, J.A. (eds.) Handbook of Logic in Artificial Intelligence and Logic Programming, vol. 3, pp. 25–110. Oxford University Press, Oxford (1994)
120. Marek, W., Shvartz, G., Truszczynski, M.: Modal nonmonotonic logics: ranges, characterization, computation. In: Proceedings of KR'91, pp. 395–404 (1991)
121. Mendelson, E.: Introduction to Mathematical Logic, 3rd edn. Wadsworth and Brooks, Monterey (1987)
122. McCarthy, J.: Circumscription—a form of non-monotonic reasoning. Artif. Intell. **13**, 27–39 (1980)
123. McCarthy, J.: Applications of circumscription to formalizing commonsense reasoning. Artif. Intell. **28**, 89–116 (1984)
124. McDermott, D.: Nonmonotonic logic II. J. ACM **29**, 33–57 (1982)

125. McDermott, D., Doyle, J.: Non-monotonic logic I. Artif. Intell. **13**, 41–72 (1980)
126. Minker, J.: On indefinite deductive databases and the closed world assumption. In: Loveland, D. (ed.), Proceedings of the 6th International Conference on Automated Deduction, pp. 292–308. Springer, Berlin (1982)
127. Minsky, M.: A framework for representing knowledge. In: Haugeland, J. (ed.), Mind-Design, pp. 95–128. MIT Press, Cambridge, Mass, (1975)
128. Miyamoto, S., Murai, T., Kudo, Y.: A family of polymodal systems and its application to generalized possibility measure and multi-rough sets. JACIII **10**, 625–632 (2006)
129. Moore, R.: Possible-world semantics for autoepistemic logic. In: Proceedings of AAAI Non-Monotonic Reasoning Workshop, pp. 344–354 (1984)
130. Moore, R.: Semantical considerations on nonmonotonic logic. Artif. Intell. **25**, 75–94 (1985)
131. Munkers, J.: Topology, 2nd edn. Prentice Hall, Upper Saddle River, NJ (2000)
132. Murai, T., Kudo, Y., Akama, S.: Towards a foundation of Kansei representation in human reasoning. Kansei Eng. Int. **6**, 41–46 (2006)
133. Murai, T., Miyakoshi, M., Shinmbo, M.: Measure-based semantics for modal logic. In: Lowen, R., Roubens, M. (eds.) Fuzzy Logic: State of the Arts. pp. 395–405. Kluwer, Dordrecht (1993)
134. Murai, T., Miyamoto, S., Kudo, Y.: A logical representation of images by means of multi-rough sets for Kansei image retrieval. In: Proceedings of RSKT 200, pp. 244–251. Springer, Heidelberg (2007)
135. Murai, T., Miyakoshi, M., Shimbo, M.: Soundness and completeness theorems between the Dempster-Shafer theory and logic of belief. In: Proceedings of the 3rd FUZZ-IEEE on World Congress on Computational Intelligence (WCCI), pp. 855–858 (1994)
136. Murai, T., Miyakoshi, M. and Shinmbo, M.: A logical foundation of graded modal operators defined by fuzzy measures. In: Proceedings of the 4th FUZZ-IEEE, pp. 151–156 (1995). (Semantics for modal logic, Fuzzy Logic: State of the Arts, pp. 395–405. Kluwer, Dordrecht (1993))
137. Murai, T., Nakata, M., Sato: A note on filtration and granular resoning. In: Terano et al. (eds.) New Frontiers in Artificial Intelligence, LNAI 2253, pp. 385–389 (2001)
138. Murai, T., Resconi, G., Nakata, M. and Sato, Y.: Operations of zooming in and out on possible worlds for semantic fields. In: Damiani, E., et al. (ed.) Knowledge-Based Intelligent Information Engineering Systems and Allied Technology, pp. 1083–1087. IOS Pres (2002)
139. Murai, T. and Sato,Y.: Association rules from a point of view of modal logic and rough sets. In: Proceeding 4th AFSS, pp. 427–432 (2000)
140. Murai, T., Nakata,M., and Sato, Y.: A note on conditional logic and association rules. In: Terano, T., et al.(ed.) New Frontiers in Artificial Intelligence, LNAI 2253, pp. 390–394. Springer, Berlin (2001)

141. Murai, T., Nakata, M., and Sato, Y.: Association rules as relative modal sentences based on conditional probability. Commun. Inst. Inf. Comput. Mach. **5**, 73–76 (2002)
142. Murai, T., Sato, Y., Kudo,Y.: Paraconsistency and neighborhood models in modal logic. In: Proceedings of the 7th World Multiconference on Systemics, Cybernetics and Informatics, vol. XII, pp. 220–223 (2003)
143. Nakamura, A., Gao, J.: A logic for fuzzy data analysis. Fuzzy Sets Syst. **39**, 127–132 (1991)
144. Negoita, C., Ralescu, D.: Applications of Fuzzy Sets to Systems Analysis. Wiley, New York (1975)
145. Nelson, D.: Constructible falsity. J. Symb. Log. **14**, 16–26 (1949)
146. Nelson, D.: Negation and separation of concepts in constructive systems. In: Heyting, A. (ed.) Constructivity in Mathematics, pp. 208–225. North-Holland, Amsterdam (1959)
147. Ore, O.: Galois connexion. Trans. Am. Math. Soc. **33**, 493–513 (1944)
148. Orlowska, E.: Kripke models with relative accessibility relations and their applications to inferences from incomplete information. In: Mirkowska, G., Rasiowa, H. (eds.) Mathematical Problems in Computation Theory, pp. 327–337. Polish Scientific Publishers, Warsaw (1987)
149. Orlowska, E.: Logical aspects of learning concepts. Int. J. Approx. Reason. **2**, 349–364 (1988)
150. Orlowska, E.: Logic for reasoning about knowledge. Z. für Math. Log. und Grund. der Math. **35**, 559–572 (1989)
151. Orlowska, E.: Kripke semantics for knowledge representation logics. Stud. Log. **49**, 255–272 (1990)
152. Orlowska, E., Pawlak, Z.: Representation of nondeterministic information. Theor. Comput. Sci. **29**, 27–39 (1984)
153. Pagliani, P.: Rough sets and Nelson algebras. Fundam. Math. **27**, 205–219 (1996)
154. Pagliani, P., Intrinsic co-Heyting boundaries and information incompleteness in rough set analysis. In: Polkowski, L., Skowron, A. (eds.) Rough Sets and Current Trends in Computing, pp. 123–130. Springer, Berlin (1998)
155. Pal, K., Shanker, B., Mitra, P.: Granular computing, rough entropy and object extraction. Pattern Recognit. Lett. **26**, 2509–2517 (2005)
156. Pawlak, P.: Information systems: theoretical foundations. Inf. Syst. **6**, 205–218 (1981)
157. Pawlak, P.: Rough sets. Int. J. Comput. Inf. Sci. **11**, 341–356 (1982)
158. Pawlak, P.: Rough Sets: Theoretical Aspects of Reasoning about Data. Kluwer, Dordrecht (1991)
159. Peirce, C: Collected Papers of Charles Sanders Peirce, vol. 8. In: Hartshone, C., Weiss, P., Burks, A. (eds.). Harvard University Press, Cambridge, MA (1931–1936)
160. Polkowski, L.: Rough Sets: Mathematical Foundations. Pysica-Verlag, Berlin (2002)

161. Pomykala, J., Pomykala, J.A.: The stone algebra of rough sets. Bull. Pol. Acad. Sci. Math. **36**, 495–508 (1988)
162. Priest, G.: Logic of paradox. J. Philos. Log. **8**, 219–241 (1979)
163. Priest, G.: Paraconsistent logic. In: Gabbay, D., Guenthner, F. (eds.) Handbook of Philosophical Logic, 2nd edn, pp. 287–393. Kluwer, Dordrecht (2002)
164. Priest, G.: In Contradiction: A Study of the Transconsistent, 2nd edn. Oxford University Press, Oxford (2006)
165. Quafafou, M.: $\alpha$-RST: a generalizations of rough set theory. Inf. Sci. **124**, 301–316
166. Rasiowa, H.: An Algebraic Approach to Non-Classical Logics. North-Holland, Amsterdam (1974)
167. Reiter, R.: On closed world data bases. In: Gallaire, H., Minker, J. (eds.) Logic Data Bases, pp. 55–76. Plenum Press, New York (1978)
168. Reiter, R.: A logic for default reasoning. Artif. Intell. **13**, 81–132 (1980)
169. Robinson, J.A.: A machine-oriented logic based on the resolution principle. J. ACM **12**, 23–41 (1965)
170. Routley, R., Plumwood, V., Meyer, R.K., Brady, R.: Relevant Logics and Their Rivals, vol. 1. Ridgeview, Atascadero (1982)
171. Sendlewski, A.: Nelson algebras through Heyting ones I. Stud. Log. **49**, 105–126 (1990)
172. Segerberg, K.: Irrevocable belief revision in dynamic doxastic logic. Notre Dame J. Form. Log. **39**, 287–306 (1998)
173. Shafer, G.: A Mathematical Theory of Evidence. Princeton University Press, Princeton (1976)
174. Shen, Y., Wang, F.: Variable precision rough set model over two universes and its properties. Soft. Comput. **15**, 557–567 (2011)
175. Shoham, Y.: A semantical approach to nonmonotonic logics. Proc. Log. Comput. Sci., 275–279 (1987)
176. Slowinski, R., Greco, S., Matarazzo, B.: Rough sets and decision making. In: Meyers, R. (ed.) Encyclopedia of Complexity and Systems Science, pp. 7753–7787. Springer, Heidelberg (2009)
177. Stalnaker, R.: A note on non-monotonic modal logic. Artif. Intell. **64**, 183–1963 (1993)
178. Subrahmanian, V.: On the semantics of quantitative logic programs. In: Proceedings of the 4th IEEE Symposium on Logic Programming, pp. 173–182 (1987)
179. Stoy, J.: Denotational Semantics: The Scott-Strachey Approach to Programming Language Theory. MIT Press, Cambridge Mass (1977)
180. Tsumoto, S.: Modelling medical diagnostic rules based on rough sets. In: Rough Sets and Current Trends in Computing, pp. 475–482. (1998)
181. Yao, Y., Lin, T.: Generalization of rough sets using modal logics. Intell. Autom. Soft Comput. **2**, 103–120 (1996)
182. van Emden, M., Kowalski, R.: The semantics of predicate logic as a programming language. J. ACM **23**, 733–742 (1976)

183. Van Gelder, A., Ross, K., Schipf, J.: The well-founded semantics for general logic programs. J. ACM **38**, 620–650 (1991)
184. Vakarelov, D.: Notes on constructive logic with strong negation. Stud. Log. **36**, 110–125 (1977)
185. Vakarelov, D.: Abstract characterization of some knowledge representation systems and the logic $NIL$ of nondeterministic information. In: Skordev, D. (ed.) Mathematical Logic and Applications. Plenum Press, New York (1987)
186. Vakarelov, D.: Modal logics for knowledge representation systems. Theor. Comput. Sci. **90**, 433–456 (1991)
187. Vakarelov, D.: A modal logic for similarity relations in Pawlak knowledge representation systems. Stud. Log. **55**, 205–228 (1995)
188. Vasil'ev, N.A.: Imaginary Logic. Nauka, Moscow (1989). (in Russian)
189. Wansing, H.: The Logic of Information Structures. Springer, Berlin (1993)
190. Wong, S., Ziarko, W.: Comparison of the probabilistic approximate classification and the fuzzy set model. Fuzzy Sets Syst. **21**, 357–362 (1987)
191. Zadeh, L.: Fuzzy sets. Inf. Control **8**, 338–353 (1965)
192. Zadeh, L.: Fuzzy sets as a basis for a theory of possibility. Fuzzy Sets Syst. **1**, 3–28 (1976)
193. Ziarko, W.: Variable precision rough set model. J. Comput. Syst. Sci. **46**, 39–59 (1993)

# Index




S. Akama et al., *Reasoning with Rough Sets*, Intelligent Systems Reference Library 142, https://doi.org/10.1007/978-3-319-72691-5

Printed in the United States
By Bookmasters

Printed in the United States
By Bookmasters